Advances in Soil Science

SUBSOIL MANAGEMENT TECHNIQUES

Edited by

N.S. Jayawardane
B.A. Stewart

LEWIS PUBLISHERS
Boca Raton Ann Arbor London Tokyo

Library of Congress Cataloging-in-Publication Data

Subsoil management techniques/ edited by N.S. Jayawardane, B. A. Stewart.
p. cm. — (Advances in soil science)
Includes bibliographical references and index.
ISBN 1-56670-020-5
1. Subsoils. 2. Soil management.
I. Jayawardane, N.S. (Nihal S.) II. Stewart, B. A. (Bobby Alton), 1932– .
III. Series: Advances in soil science (Boca Raton, Fla.)
S593.27.S83 1994
631.4—dc20 94-28050
CIP

International Standard Book Number 1-56670-020-5
Library of Congress Card Number 94-28050
Printed in the United States of America 1 2 3 4 5 6 7 8 9 0
Printed on acid-free paper

About the Editors:

Dr. N.S. Jayawardane is a Research Scientist in the CSIRO, Division of Water Resources, Australia. He has had responsibility for developing a novel technique for improving sodic and acid subsoils. He is currently developing a new technique for waste effluent management through land filtration combined with intensive cropping. Before joining CSIRO in 1982, Dr. Jayawardane worked in the Land Use Division of the Irrigation Department of Sri Lanka, carrying out soil surveys for agricultural projects. He was responsible for carrying out detailed soil mapping of the Wet zone, which was incorporated into the revised Soil Map of Sri Lanka. Dr. Jayawardane has served as General Secretary and Editor of the Soil Science Society of Sri Lanka.

Dr. B.A. Stewart is a Distinguished Professor of Soil Science, and Director of the Dryland Agriculture Institute at West Texas A&M University, Canyon, Texas. Prior to joining West Texas A&M University in 1993, he was Director of the USDA Conservation and Production Research Laboratory, Bushland, Texas. Dr. Stewart is past president of the Soil Science Society of America, and was a member of the 1990-93 Committee of Long Range Soil and Water Policy, National Research Council, National Academy of Sciences. He is a Fellow of the Soil Science Society of America, American Society of Agronomy, Soil and Water Conservation Society, a recipient of the USDA Superior Service Award.

Contributors

J. Blackwell, CSIRO, Division of Water Resources, Private Mail Bag No. 3, Griffith, NSW 2680, Australia.

B. Cockroft, Institute of Sustainable Irrigated Agriculture, Department of Agriculture, Private Bag, Ferguson Road, Tatura, VIC 3616, Australia.

R. Horn, Institut für Pflanzenernährung und Bodenkunde, Christian-Albrechts-Universität Zu Kiel, Olshausenstraβe 40, D-2300 Kiel 1, Germany.

N.S. Jayawardane, CSIRO, Division of Water Resources, Private Mail Bag No. 3, Griffith, NSW 2680, Australia.

G. Kirchhof, Department of Agriculture, University of Queensland, St. Lucia, QLD 4067, Australia.

R. Lal, Department of Agronomy, The Ohio State University, Columbus, OH 43210-1086.

K.E. Lee, CSIRO, Division of Soils, Private Bag No. 2, Glen Osmond, SA 5064, Australia.

W.A. Muirhead, CSIRO, Division of Water Resources, Private Mail Bag No. 3, Griffith, NSW 2680, Australia.

K.A. Olsson, Institute of Sustainable Irrigated Agriculture, Department of Agriculture, Private Bag, Ferguson Road, Tatura, VIC 3616, Australia.

P. Rengasamy, Institute of Sustainable Irrigated Agriculture, Department of Agriculture, Private Bag, Ferguson Road, Tatura, VIC 3616, Australia.

K.R.J. Smettem, CSIRO, Division of Soils, Davies Laboratory, Townsville, QLD 4814, Australia.

Gordon Spoor, Silsoe College, Cranfield Institute of Technology, Silsoe Bedford MK45 4DT, England

Malcolm E. Sumner, Department of Crop and Soil Sciences, The University of Georgia, Athens, GA 30602-7272.

Preface

Subsoil problems that reduce crop productivity include sodicity, salinity, acidity, low fertility, compaction, and waterlogging. Extension of agriculture to soils of low productivity with adverse subsoil properties has been necessitated by increased demands for food and clothing for the expanding world human population. On existing agricultural lands, inadequate management practices such as excessive cultivation and trafficking with heavy farm equipment, indiscriminate fertilizer use and poor irrigation and agronomic management practices have contributed to the development of these subsoil problems.

Traditional tillage systems for correcting these adverse subsoil conditions usually involve massive soil disturbance through deep plowing or ripping, sometimes combined with chemical ameliorant additions. However, these techniques have often failed to provide long-term, sustainable improvements of the soil for crop production. This is often due to poor mixing of soil ameliorants to depths and difficulties associated with maintaining the improved soil structures created by massive soil loosening, especially under trafficking and flood irrigation. These tillage techniques have sometimes also led to other problems such as soil erosion and hard pan formation.

This publication reviews alternative approaches to overcoming subsoil problems. These approaches involve reduced disturbance of the soil, but still provide substantial and sustainable soil improvements. Thus, in Chapters 1 to 4 of this book, the use of minimum tillage, bed farming, mole drainage and slotting to overcome adverse subsoil conditions is discussed. Chapter 5 examines the options available for management of subsoil acidity. Economic viability of these improved subsoil amelioration techniques often depend on the subsequent management practices adopted to conserve and stabilize such improvements, especially in maintaining improved soil structure. This requires an understanding of the processes involved in stress transmission and compaction under farm trafficking (Chapter 6) and the critical role of soil microorganisms in providing long-term biological stabilization and improvements of soils (Chapter 7).

The best approach to address subsoil deficiency problems in a given location depends on the specific local soil, climatic, cropping, and economic factors. Soil improvement to provide both initial and long-term benefits may require combinations of the soil ameliorative techniques described above. It is hoped that this publication, which brings together the latest information on these ameliorative techniques, could be a useful guide to land managers and researchers attempting to find optimum solutions to the specific subsoil problems in their region. In order to assist in this process, the review papers have been written with emphasis on developing new ideas and innovative adaptations and on research to find practical solutions to field problems.

N.S. Jayawardane
Griffith, Australia

B.A. Stewart
Amarillo, Texas

Contents

Minimum Tillage Systems

R. Lal

I. Introduction

Suboptimal subsoil conditions have significant impacts on soil-water and nutrient regime, gaseous exchange between soil and the atmosphere, and crop growth. Adverse effects on crop growth have agronomic and economic implications, while those on soil-water and aeration regimes lead to ecological and environmental problems of regional and global importance (Figure 1). These problems may be edaphological leading to reduced crop/animal production, and loss of inputs or high inputs required to produce the desired level of output and services. Edaphological impacts of adverse subsoil conditions are generally local or

1-56670-020-5/95/$0.00+.50

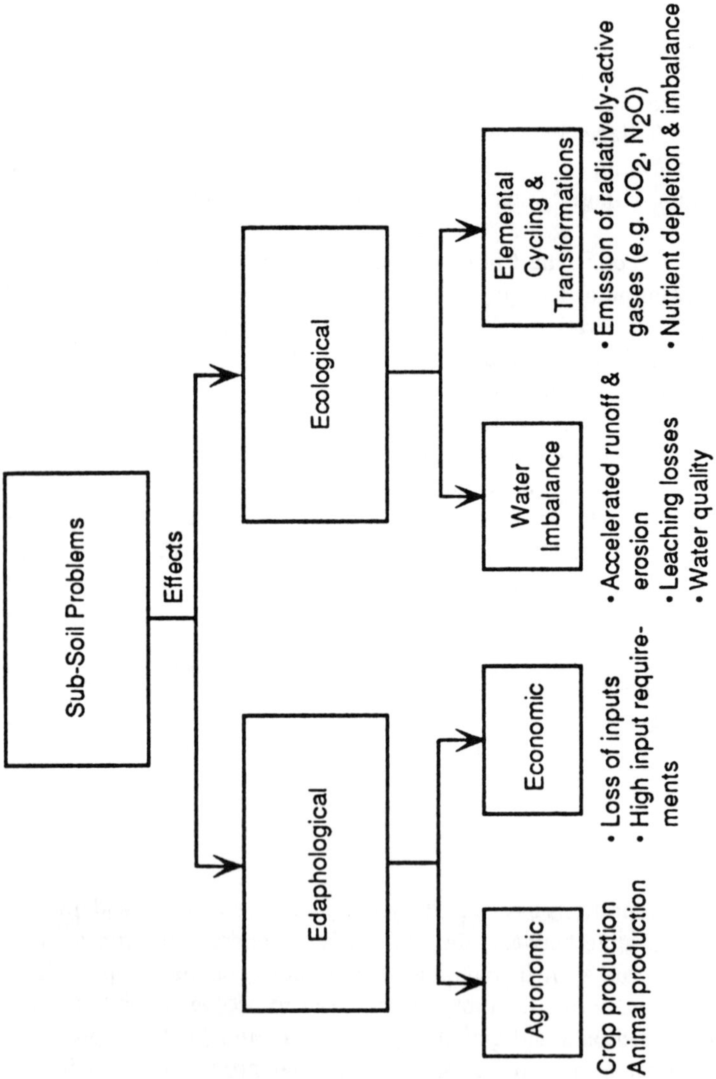

Figure 1. Edaphological and ecological effects of suboptimal subsoil conditions.

regional, and their magnitude depends on soil and environmental attributes, and the characteristics of cropping/farming systems used.

Ecological effects of subsoil problems are regional and global in nature. Principal among these are related to water quality and emission of radiatively-active gases from soil to the atmosphere. Water quality is influenced by perturbations in components of the water balance, e.g., increase in surface runoff, accelerated erosion, transport of sediments and sediment-borne contaminants and dissolved chemicals in surface runoff and seepage flow, macropore flow and chemical transport with macropore flow. Globally, accelerated erosion by water may lead to annual displacement of about 140 to 190 Pg (Pg = petagram = 1×10^{15} grams) of soil over the landscape (Lal, 1993). Reduction in agricultural productivity on-site, and pollution of natural water and damage to waterways and civil structures off-site are among principal effects of accelerated soil erosion.

World soils can be a major source or sink of radiatively-active gases. Emission of CO_2, CH_4 and N_2O from world soils are estimated at 1173, 60 and 4.7 Tg (Tg = teragram = 1×10^{12} g) per year (Watson, et al., 1992). Total emissions of these gases from agricultural sources in 1990 were estimated at 6527, 198 and 5.3 Tg for CO_2, CH_4 and N_2O, respectively (Watson, et al., 1992). Soil surface management in general, and tillage practices in particular, play a major role in magnitude and seasonal trends in gaseous emission from soils. In contrast to emissions, judicious and discriminate use of tillage systems can lead to carbon sequestration in soil.

Minimum tillage, therefore, is an ecological approach to resource management and sustainable agriculture. It is a science-based strategy to intensify agricultural production without causing serious degradation of soil resource and with a minimum damage to the environment. Despite its potentially positive effects on soil and environment, minimum tillage is a practical approach to alleviate specific subsoil related constraints to crop production.

II. Types of Minimum Tillage Systems

The magnitude of soil disturbance, and its susceptibility to degradative processes, depends on several factors including frequency and intensity of tillage, type of tillage implements, source of power, and antecedent soil properties. Soil's susceptibility to tillage-induced degradative processes can be drastically decreased by reducing frequency and intensity of tillage operations, choosing those tillage implements that loosen the soil without turning over and do not excessively pulverize it, and by performing the needed tillage operations when soil moisture conditions are within the optimum range to produce the desired tilth. These are essentially the characteristics/attributes of a "minimum tillage" system.

Minimum tillage is a generic term and refers to any tillage sequence based on "minimum soil manipulation necessary for crop production or meeting tillage requirements under the existing soil and climatic conditions" (Resource Conservation Glossary, 1982). The term minimum tillage is often used synonymously with conservation tillage. The latter is defined "any tillage sequence that reduces loss of soil or water relative to conventional tillage; often a form of non-inversion tillage that retains a protective amount of mulch on the surface" (Resource Conservation Glossary, 1982). Principal criteria in minimum tillage or conservation tillage systems are: (i) reduction in frequency or intensity of tillage, (ii) presence of crop residue mulch on the soil surface, and (iii) increase in random roughness or surface water detention capacity. These criteria can be met by a broad range of tillage systems outlined in Figure 2.

Most of the minimum tillage systems outlined in Figure 2 are used interchangeably under the broad umbrella term of "conservation tillage". Some commonly used minimum tillage systems are briefly described below:

A. Reduced Tillage Systems

A broad range of reduced tillage systems are used to conserve soil and water, and enhance and sustain agricultural productivity. Important types of reduced tillage systems include the following:

(i) No-till: It is a tillage system whereby crop is sown directly into an untilled seedbed without any primary or secondary tillage. Previous crop residue is left on the surface, and weeds are generally controlled by herbicides. This system is also called zero-tillage or no-tillage.

(ii) Slit or Slot Tillage: A narrow slot is opened to facilitate the seed/soil contact for good germination and satisfactory stand establishment. Similar to the no-till system, previous crop residue is left on the soil surface as mulch and weeds are controlled by herbicides.

(iii) Chisel tillage: Soils prone to subsoil compaction are loosened to 30 cm. or 50 cm. depth by chisel tillage done without substantial soil turnover. Crops are sown directly above the chiselled row.

(iv) Plow plant: Plow based inversion tillage is used to break the compacted soil surface without additional secondary or tertiary tillage operations. Plowing is generally done during fall (fall tillage) in temperate climate, and at the end of the rainy season in tropical regions.

(v) Disc plant: Disc plow rather than moldboard plow is used to break the surface crust and leave maximum crop residue on the soil surface. Disc plowing is also done in the fall, followed by light harrowing in the spring just before sowing.

(vi) Subsurface tillage: These tillage systems are similar to chisel tillage, and are used to alleviate subsoil compaction or pans, e.g., clay pan, plow pan, etc. Subsurface tillage involves little or no soil inversion. Commonly used tools for subsurface tillage are paraplow or rod weeder.

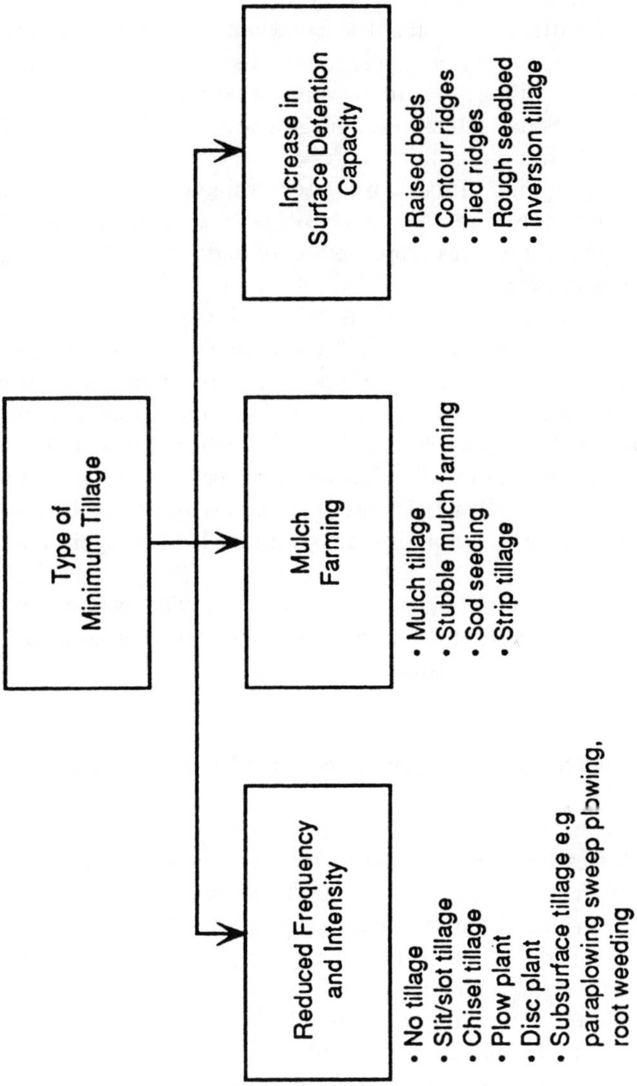

Figure 2. Principal types of minimum tillage systems.

B. Mulch Farming

A minimum tillage system that ensures a maximum retention of crop residue on the soil surface is called *mulch tillage* or, *stubble farming* or *mulch farming*. It is defined as preparation of the soil in such a way that plant residue or other mulching materials are specifically left on or near the surface. Mulch tillage is also a broad term and includes practices such as no-tillage, disk plant system, chisel tillage, and subsurface tillage that leave maximum amount of crop residue on the soil surface. Principal types of tillage systems under this category are:

(i) Mulch tillage: Any tillage system in which at least 30% of the ground is covered by crop residue mulch. Percent ground covered by the residue mulch is an import criterion in this category.

(ii) Stubble mulch farming: This term is used synonymously with mulch tillage. The amount of crop residue mulch available at the time of sowing depends on the type of previous crop, method of harvest, residue management system, grazing, etc.

(iii) Sod seeding: When a grain crop is seeded through the mulch of a chemically killed cover crop, it is called "sod seeding". If the cover crop is either not killed or is only temporarily suppressed, it is called "live mulch". A system of growing a cover crop to restore soil structure, replenish plant nutrient reserves, and control weeds and pests is called "planted fallow", "ecofallow" or "summer fallow". Ecofallow is practiced during the summer in arid climates when cropless fields are occasionally disced or harrowed to control weeds, conserve soil moisture, and protection soil against wind erosion.

(iv) Strip tillage: A narrow strip, usually 10 to 15 cm. wide, is prepared by mechanical manipulation through the sod or live mulch system to facilitate seed/soil contact and obtain satisfactory crop stand.

C. Minimum Tillage Systems to Increase Surface Detention Capacity

When infiltration capacity of the subsoil is restrictively low, it is appropriate to minimize runoff losses by increasing the surface detention capacity and allow more time for water to soak into the ground. Different types of minimum tillage systems to increase surface detention capacity include the following:

(i) Raised beds: This refers to a method of seedbed preparation whereby the topsoil is scraped and concentrated in a defined region to deliberately raise the seedbed above the natural terrain. Raised beds involve soil disturbance in only part of the field from where the soil is scraped or removed. There are several types of raised beds depending on the objectives and mode of construction. Raised beds may be constructed by digging drainage furrows at regular intervals and putting the dugout soil in between the furrows to raise the seedbed where crops are sown. Raised beds may be mounds or hillocks widely used in West Africa, primarily for growing root crops within a multiple cropping system. Large mounds are constructed in wetlands and hydromorphic soils to facilitate growth of upland crops on the mound and aquatic crops in between the mounds. These methods of seedbed preparation are traditionally performed by manual operations with simple tools, e.g., hoe, digger, etc. Raised beds are cultural traditions in many ancient agricultural societies (Lal, 1990).

(ii) Contour ridges: The practice of planting or seeding crops on ridges is a widely adapted system in both temperate and tropical climates. The crop row may be planted on the ridgetop, along both ridge sides, or in the furrow. Ridge tillage facilitates the mixed cropping systems - a practice of growing more than one crop simultaneously in the same plot of land, a system commonly practiced throughout the tropics an subtropics. The ridges may be made every season. Alternatively, in a semipermanent ridge-furrow system, only a necessary repair is done at the onset of a new cropping cycle. Ridges are generally built on the contour with graded furrow draining into a channel or grass waterway (Plate 11). Merits and limitations of ridge tillage systems are outlined in Table 1.

(iii) Tied ridges: Stability of contour ridges can be enhanced by constructing short cross-ties to create a series of basins to store water. The system with cross-ties is called the tied-ridge system or basin tillage (Plate 12).

(iv) Rough seedbed: A system of rough plowing, done at the end of the rainy season to deliberately create large clods and an uneven soil surface, is adopted to increase surface detention capacity and protect the soil against wind erosion during the dry period. This system is recommended for sandy soils of the West African Sahel (Chopart and Nicou, 1989).

(v) Inversion tillage: This tillage system is similar to the rough seedbed, and is done as a primary tillage using moldboard plow at anytime when there is a need to protect the soil against erosion. Inversion tillage done to control erosion is also called "emergency tillage".

D. Factors Affecting the Choice of Minimum Tillage Type

The minimum tillage (e.g., intensity, frequency, and type) required to successfully grow a crop and effectively conserve soil and water resources depend on several factors (Figure 3). Among biophysical factors, soil and climatic factors are the most critical. Soil and climatic factors can be lumped together into "land attributes" which in addition, also implies terrain characteristics, climax vegetation, and drainage density. Important among terrain characteristics in relation to choice of a minimum tillage system are slope gradient, slope length, and slope shape. Texture, structure and drainage are important soil factors, whereas length of the growing season, climatic erosivity, and water balance are relevant climatic factors. Farm size, land tenure rights, and institutional factors are pertinent socio-economic and cultural factors.

Table 1. Merits and demerits of ridge tillage

Merits	Demerits
Saving in labor	Compaction in the furrows
Erosion control	Drought on ridges
Traffic control	High cost
Enhancing soil fertility	Erosion
Precision planting	Seedling submergence
Control of gravity flow irrigation	High temperature fluctuations
Drainage of excess water	Crop lodging
Crop establishment on saline soils	
Moisture conservation	
Early sowing	
Multiple cropping	
Inter-cultivation	
Facilitating weed control	
Easy harvest of root crops	
Improving water management	
Reducing adverse effects of sand blasting	

III. Need for Minimum Tillage

Plow-based mechanical soil disturbance, although necessary for some soils for an optimum seedbed preparation and for weed control, has many ecological and agronomic limitations including: (i) high susceptibility of soil to erosion, (ii) rapid mineralization of soil organic matter leading to emission of radiatively-active gases to the atmosphere, (iii) subsoil compaction and formation of plow pan, and (iv) high energy costs. Conventional tillage, based on primary and secondary tillage operations for soil turnover and pulverization, despite transient beneficial effects, may accentuate problems of soil and environmental degradation. Yet crops have to grow with a minimal competition from weeds and expensive inputs have to be economized and efficiently used. Minimum tillage system is a compromise strategy that facilitates intensive cultivation with minimal risks of soil and environmental degradation. However, these systems are not without their own limitations. Principal advantages and limitations of minimum tillage systems are listed in Table 2. For achieving the maximum benefits from potential advantages, however, the limitations of minimum tillage systems must be carefully addressed. Most limitations, similar to advantages, are soil and crop specific. Addressing these limitations requires adaptive research based on system-oriented and holistic approach.

While minimum tillage systems have specific advantages of addressing subsoil problems, two specific advantages of these systems are energy conservation and sustainable use of soil and water resources.

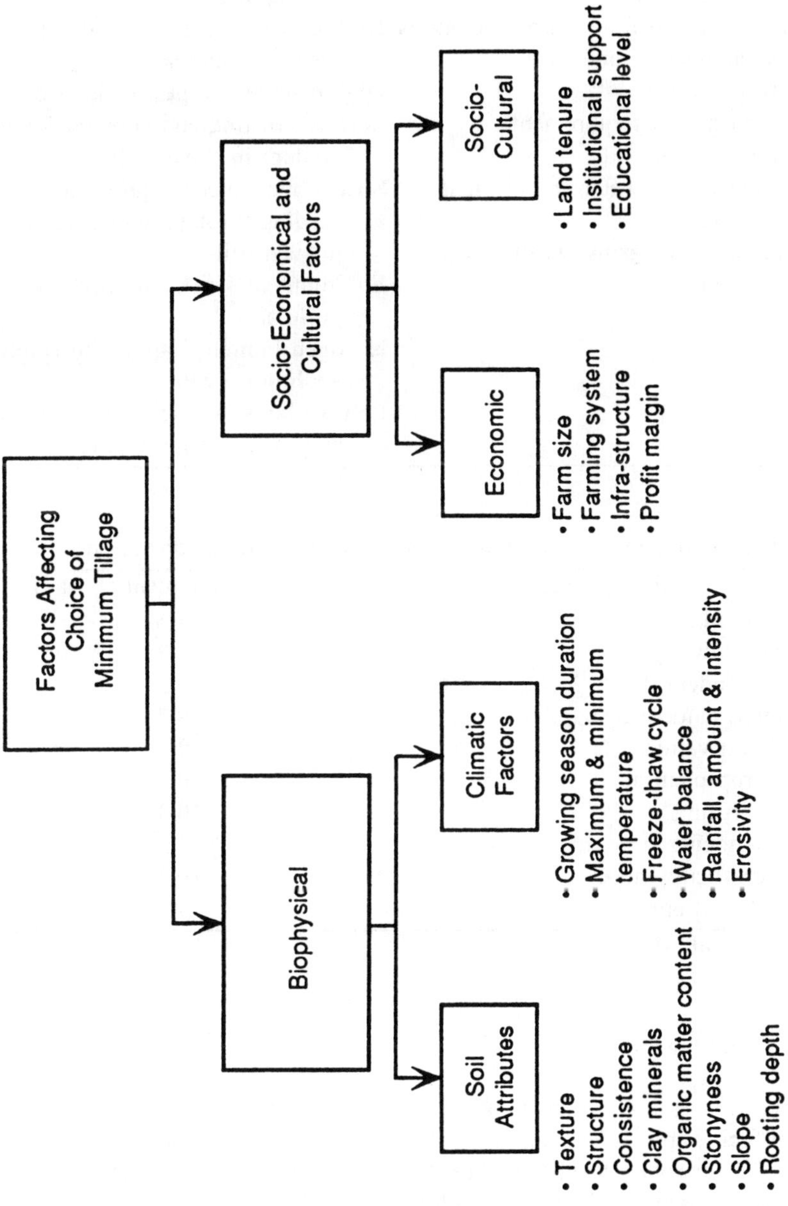

Figure 3. Factors affecting the choice of minimum tillage for sustainable use of soil resources

Table 2. General advantages and limitations of minimum tillage systems

Advantages	Disadvantages
Saving time for seedbed preparation	Stubble disposal in some environments
Saving in fuel cost	High cost of herbicides and heavy weed infestation
Less traffic and fewer farm operations	High disease and pest incidence
Early land access and easy trafficability	Surface crusting and poor seed-soil content in some soils
Less machinery requirement	Special machinery requirements for seeding through mulch on an uneven soil
Less erosion hazard	Environmental risks of chemical pollutants
Efficient use of water and nutrient reserves	Nitrogen immobilization and stunted seedling growth
Yield stability against weather, e.g., drought	Lower yields on poorly drained soils and in favorable weather

Table 3. Estimate of fuel requirements of various tillage operations

Field Operation	Fuel Requirement (L ha^{-1})
Moldboard plowing	21.6
Chisel plowing	14.3
Heavy tandem disk	12.2
Standard tandem disk	9.7
Field cultivation	9.7
Row crop planting	10.3
Grain drilling	10.3
Rotary hoeing	7.3
Row crop cultivation	10.3
Knifed NH_3 application	13.0

(From Stout, 1984.)

A. Energy Conservation

Conventional tillage is an energy-intensive activity, and is highly dependent on fossil fuel (Table 3, Ritchie, 1983). Fuel requirements in liters per hectare range from 7.3 for rotary hoeing, to 21.6 for moldboard plowing. In comparison, the energy requirement for a no-till planter is 10.3 liters per hectare. In the U.S. primary and secondary tillage operations consume 10-12 trillion Kcals a year (Stout, 1984). Field operations consume 10 to 15 percent of the total energy input, and most of the energy used is non-renewable fossil fuel. Elimination of moldboard plowing, therefore, can lead to a considerable savings in fuel consumption. However, herbicides used in minimum tillage systems are also based on fossil fuel (Stout, 1984) and require additional energy inputs.

B. Sustainable Use of Soil and Water Resources

Sustainability implies profitable farming on a continuous basis while enhancing the productive capacity of soil and water resources. However, sustainable agriculture must look beyond production economics to ecological stability and environmental quality. Minimum tillage systems affect all aspects of agricultural sustainability (Figure 4). Sustainable use of soil resources is directly impacted by tillage-induced changes in several key soil processes, e.g., soil structure and its improvement, erosion control, carbon sequestration and soil organic matter enhancement, water conservation and efficient use of water resources, aeration, etc. Major environmental aspects of minimum tillage system are related to emission of radiatively-active gases (e.g., CO_2, CH_4 and N_2O) from soil and to dust concentration in air by wind erosion control. Similar to air quality, water quality is impacted by tillage-related impact on transport of sediments and sediment-laden pollutants. Transport of agricultural chemicals through the soil is another important aspect of water quality. Productivity, especially with regards to long-term trends, is related to yield stability and production per unit area and time, and with regard to consumption of the most limited resource, e.g., fossil fuel, decline in soil organic matter content, etc.

Minimum tillage systems, because of several limitations listed in Table 1, may not always have positive impacts on sustainability. The objective of adaptive research is to minimize those limitations, produce synergistic effects, and enhance sustainability.

IV. Managing Subsoil Compaction

Subsoil compaction is a serious problem, especially in mechanized intensive agriculture. However, total economic impact of soil compaction is difficult to assess. On-farm losses due to compaction-related yield reduction in the U.S. may be in excess of one billion dollars annually (Raghavan, et al., 1990). In Russia, Rusanov (1991) observed yield reduction in grain and fodder production due to subsoil compaction in crops ranging from 7.8 to 35.1% compared with an uncompacted control. The magnitude of yield reduction by soil compaction may be higher in tropical than in temperate region soils.

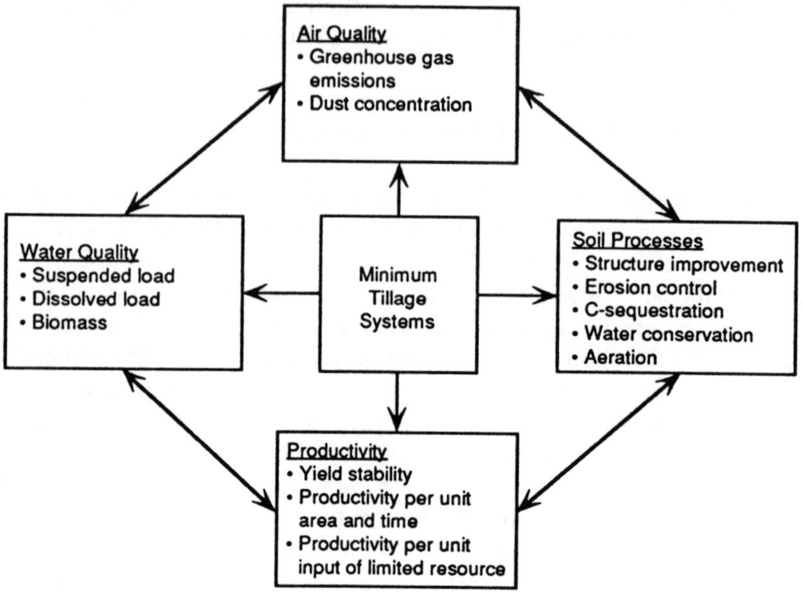

Figure 4. Minimum tillage systems effects on sustainability in relation to resource use, environmental quality, and productivity.

A. Preventing Subsoil Compaction

Minimum tillage systems can play a major role in decreasing risks of subsoil compaction. The beneficial effects of minimum tillage systems are attributed to: (i) fewer farm operations, (ii) cushioning/buffering effect of crop residue mulch, (iii) high soil organic matter content, and (iv) high soil biodiversity, e.g., activity and species diversity of soil fauna.

There are several examples of low subsoil compaction with minimum tillage systems in both temperate and tropical regions. In Mato Grosso, Brazil, Centurion (1987) observed that surface soil was compacted in minimum tillage systems and subsoil was compacted in conventional tillage systems. Compacted soil layers were observed at 3 cm. depth in no-tillage, at 10-13 cm. depth in minimum tillage, and at 20-23 cm. depth in conventional tillage and double plowing. Similar observations were made in Columbia, Missouri, U.S. by Gebhardt, et al. (1986). They observed that soil compaction in the top 7.5 cm. was more for no-tillage than conventional tillage. However, soil compaction at 17.8 cm. was more for conventional tillage than no-tillage. In Nigeria, Oni (1991) also observed that surface soil compaction was higher for no-tillage than other methods of seedbed preparation. The data in Table 4 for a Brazilian Oxisol

Table 4. Tillage effects on physical properties of a Brazilian Oxisol

Soil Properties	No-Tillage	Minimum Tillage	Conventional Tillage
A. 0 - 10 cm.			
Bulk density (Mg m^{-3})	1.06a	0.97b	0.99b
Total porosity (%)	61.39a	64.76b	64.69b
Macropores >50 μm (%)	22.71a	28.44b	27.72ab
B. 10 - 20 cm.			
Bulk density (Mg M^{-3})	1.15a	1.07b	1.08b
Total porosity (%)	57.72a	61.11b	61.27b
Macropores >50 μm (%)	14.14a	20.91b	20.93b

(From Roth et al., 1988.)

show that bulk density at 20-30 cm. depth was significantly lower in no-tillage and minimum tillage systems compared with the conventional tillage system. Accordingly, the total porosity was significantly higher in minimum-tillage and no-tillage compared with conventional tillage systems. Total volume of macropores (>50 μm) was also more in minimum tillage and no-tillage systems. However, soil physical properties for the surface 0-10 cm. and 10-20 cm. were not so favorable especially in no-tillage system. Soil bulk density was higher, and total macroporosity lower in no-tillage compared with conventional tillage systems. It is apparent, therefore, that conventional tillage led to development of a plowpan at 20-30 cm. depth, restricting root development and water infiltration. The data in Table 5 support the hypothesis that lack of plowpan formation and higher percent of macropores in no-till and chisel plow system lead to higher infiltration rate compared with conventional tillage. Experiments conducted on Alfisols in Australia corroborate the data obtained in Brazil. The data in Table 6 show that pore density and percent of pores transmitting the dyed water were significantly higher in no-tillage compared with conventional tillage systems. Soil porosity and percent of transmission pores in no-tillage systems were similar to that of the plots with rotational pasture. Both Brazilian and Australian data clearly show that risks of subsoil compaction can be drastically reduced by adoption of minimum tillage systems. Effectiveness of minimum tillage systems in minimizing compaction is greatly enhanced when used in association with crop residue mulch and on structurally-active soils.

Formation of plowpan is due to the smearing effect of plow sole and depends on antecedent soil properties, e.g., clay type and content, moisture content, etc. The data by Sommer and Zach (1992) from Germany also indicate the distinct formation of plowpan at 25-30 cm. depth by moldboard plowing for 7 consecutive years from 1985 to 1990 compared with no-tillage systems. The data obtained in Spring of 1990 showed that total porosity in moldboard plowed

Table 5. Tillage effects on infiltration rate (mm h^{-1}) at two moisture contents with rainfall simulator on an Oxisol in Brazil

Tillage Method	Gravimetric Moisture Content (%)	
	18 - 21 cm	30 - 34 cm
Conventional	51	48
Chisel plow	57	53
No-till	58	58
LSD (0.05)	5	9

(From Sidras and Roth, 1987.)

Table 6. Tillage and land use effects on macropores of an Australian Alfisol

Tillage Method	Pore Density (m^{-2})	% Stained
Conventional	104	8
No-tillage	186	55
Pasture	172	74
Permanent pasture	364	77
LSD (0.5)	72	23

(From Chan and Mead, 1989.)

versus no-till treatments, respectively, for 25-30 cm. depth was 33.3 vs. 35.4 percent for wheel track zone and 35.3 vs. 38.2 non-wheel track zone.

B. Alleviating Subsoil Compaction

Several minimum tillage systems can be used to alleviate subsoil compaction. Most relevant among these are chisel tillage and sub-surface tillage by equipment involving chisel plow, paraplow, etc. Loosened slits act as guidelines for roots to extend below the compacted zone, and loosen the compacted layer (Steinert, et al., 1990). Loosening by subsurface tillage is considered necessary when subsoil density exceeds 1.60 Mg/m^3. Root growth is practically zero when subsoil density exceeds 1.80 Mg/m^3 (Vepraskas, 1988). The beneficial effects of loosening, however, vary among soils (Holloway, et al., 1991).

In Germany, Sommer and Zach (1992) recommended the use of several minimum tillage systems for alleviating traffic-induced soil compaction. The data in Table 7 show that porosity below the plow depth of 20 cm. was altered by chisel tillage. Subsurface tillage operations required for alleviating subsoil compaction are expensive and energy-intensive. Small farm-holders of the tropics and several farmers even in developed economies cannot afford it. Therefore, the best overall strategy is to prevent subsoil compaction through an appropriate use of minimum tillage system.

Table 7. Porosity at field capacity (%) measured in 1990 for 25-30 cm depth for the non-wheat track group vs. chisel plow system in Germany

Cropping system	Conventional Tillage	Chisel Plowing	Significance
Sugar Beet	33.3	35.4	xx
Winter Wheat	32.9	36.7	xx
Wheat Barley	35.1	35.6	xx

xx Significantly different at 1% level of probability.
(From Sommer and Zach, 1992.)

Table 8. Tillage and compaction effects on maize grain yield (Kg ha^{-1}) in 1984 in northern Nigeria

	Compaction Level			
Tillage Method	None	Medium	Heavy	Mean
Conventional	552	5525	1350	2476
Shallow	1080	5415	1350	2608
Deep	1560	3889	48	1829
No-tillage	215	1682	42	446
Mean	852	3476	692	
LSD (0.05)				
Compaction	2261			
Tillage	1253			
C x T	3255			

(From Onwualu and Anazodo, 1989.)

Minimum tillage systems are not always effective in alleviating subsoil compaction. This is particularly true in soils of the arid and semi-arid tropics, e.g., the West African Sahel. Nicou (1977) and Charreau (1977) observed in sandy soils of the Sahel that total porosity, available water reserves, and root system development were significantly improved by deep plowing. Deep moldboard plowing and soil inversion increased infiltration, and decreased runoff and evaporation. Similar observations were made on compacted soils in Bangladesh where Rahman and Islam (1989) showed that no-tillage and shallow tillage (minimum tillage systems) were not effective in improving infiltration rate. The maximum infiltration rate was observed by deep tillage to 22.5 cm. depth. The data in Table 8 from northern Nigeria also show that conventional tillage, shallow tillage or deep tillage systems were necessary to alleviate adverse effects of soil compaction on maize grain yield. Maize grain yield in no-tillage systems was significantly lower than other tillage treatments. In Morocco, Oussible and Crookston (1987) reported that subsoiling to 0.7 m depth at 0.4 m intervals decreased soil bulk density by 11%, increased air-filled porosity by 17%

and total porosity by 50%, and increased wheat grain yield by 48% and straw yield by 28%. In Baluchistan, Pakistan, Khan and Raza (1989) observed that in comparison with shallow tillage plowing to 20 cm. depth increased soil moisture storage and crop yield. Beneficial effects of deep ripping on crop growth and yield have also been reported from Victoria, Australia (Brooke, et al., 1989).

The duration of or residual effect of the benefits of deep plowing and subsoiling depends on soil properties and subsequent management. In some soils and environments, beneficial effects are transient and short-lived. Chopart et al. (1981) observed that beneficial effects of deep primary tillage on sandy soils of the Sahel were transient in nature, and that soil structure was easily degraded during the cropping phase. Chopart and Nicou (1989) summarized the results of 20 years of continuous cropping in Senegal. They observed that although higher yields were observed with than without plowing, there was a deficiency trend in soil nutrient reserves and soil fertility. Furthermore, Costa et al. (1989) observed that crusting was increasing with frequent plowing, and runoff losses were similar in no-till and plowed soil about 6 to 7 months after plowing. Observations similar to those of the Sahelian soils were made for a hard-setting soil at Cowra, NSW, Australia. Mead and Chan (1988) observed that the effect of deep tillage on seedling vigor was short-lived. Deep-tilled plots settled to bulk density and soil strength similar to that of no-till plots. Nonetheless, it is generally agreed that structurally-inert soils of the semi-arid region require periodic deep tillage for amelioration of soil structure. Some form of minimum or reduced-tillage is necessary for sustainable management of these hard-setting and structurally inert soils.

C. Positive Effects of Soil Compaction

Sandy soils containing low organic matter content and those characterized by single-grained and loose structure can be benefitted by moderate levels of soil compaction. Because of the predominance of macropores, these soils are excessively drained, have low plant-available water reserves, and are extremely susceptible to drought. There is an optimum range of soil bulk density at which soil conditions are most favorable for crop growth. In the savanna zone of northern Nigeria, Onwualu and Avazoda (1989) observed that the optimum soil bulk density for these soils was 1.2 Mg/ m^3. Maize yield was substantially decreased when soil bulk density was below 1.2 Mg/m^3. They developed a regression equation relating maize yield with soil bulk density (Eq. 1):

$$Y = -28.65 + 52.0\rho_b - 21.8\rho_b^2, \quad R = 0.7 \qquad \textbf{(Eq. 1)}$$

where Y is maize grain yield in Mg ha^{-1} and ρ_b is soil bulk density in Mg/m^3. Light soil compaction reduces the volume and continuity of large pores, and increases relative proportion of retention pores. Nimmo and Katherine (1988) observed that light compaction of sandy soils at low water content improves water retention by the following mechanisms: (i) bridging fine particles closer together, e.g., close packing, and (ii) altering orientation of particles that influence pore size distribution. Agarwal (1991) observed for several sandy soils of the northwestern arid region that light soil compaction led to a substantial saving of irrigation water ranging from 15 to 36%. Because of low percolation losses, light compaction of sandy soil also decreases leaching losses of plant nutrients and increases crop yield.

V. Improving Water-Intake Rate

There are several strategies to increase infiltration rate of soil (Figure 5). A feasible strategy for structurally active soils is to increase volume and continuity of macropores through appropriate biotic or mechanical techniques. Biopores (e.g., root channels and macropores created by burrowing activity of earthworms and other soil fauna) are extremely important in enhancing infiltration rates. Minimum tillage systems, especially when used in conjunction with crop residue mulch, provide environments suitable for creation and maintenance of biopores. In addition to biopores, presence of crop residue mulch and any vegetative cover protects the soil surface against raindrop impact and formation of surface seal.

(i) Biotic measures: Importance of conservation tillage, mulch farming techniques, and use of vegetative measures in improving soil structure and water infiltration rate are widely recognized (Cornish and Pratley, 1987; Unger, 1990). On a Vertic soil in Maharashtra, India, Dongale (1987) observed that double cropping increased both accumulative infiltration and infiltration rate by 26 to 34% compared with monocropping. This increase in water transmission on these slowly permeable soils may be due to : (i) increased ground cover leading to buffering effect against raindrop impact, (ii) increase in shear resistance against flowing water retarding its velocity and providing more time for water to infiltrate into the soil, (iii) more biomass, especially the root biomass, to maintain soil's biotic activity, and (iv) more root channels to conduct water into the subsoil. Lack of crust formation under vegetative cover is an important factor responsible for high infiltration rate (Moran, et al., 1988).

In Parana, Brazil, Roth et al. (1986) reported higher infiltration rate under no-tillage than under conventional tillage. Similar to Brazil, considerable research information exists about application of minimum tillage systems for several ecoregions in Australia. For hard-setting soils in NSW, Chan and Mead (1988) observed that soil managed by no-till treatment had more macropores and higher infiltration rate than plowed soil. No-tillage and minimum tillage systems are useful in preserving the structure of the top 25 mm. layer of these fragile soils. Similar observations were made for a Krasnozem in North Queensland by Prove, et al. (1990).

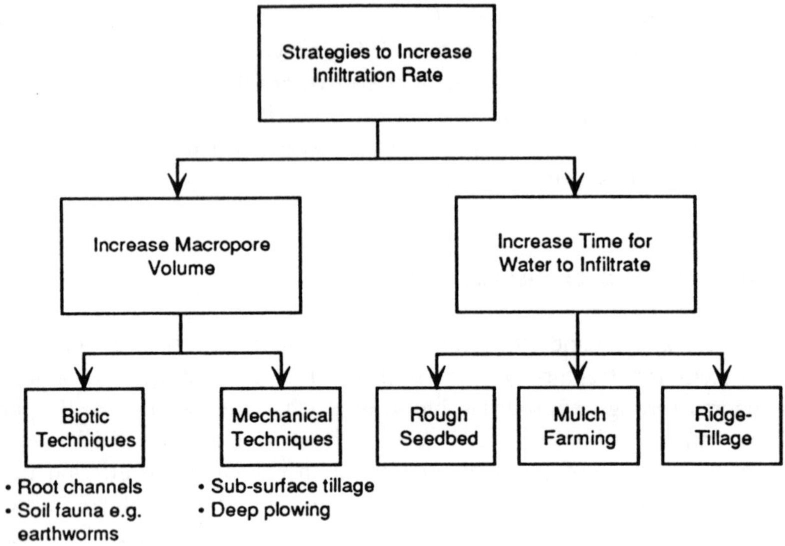

Figure 5. Strategies to enhance infiltration capacity of the soil.

They observed that no-tillage systems with stubble retention had significant and positive effects on soil physical properties and infiltration rates. In contrast, however, minimum tillage systems without crop residue mulch were not effective for a Vertisol (Thorburn, 1992). In these heavy-textured soils, minimum tillage caused severe soil degradation, especially where residue was removed. Stubble retention improved soil physical properties. In Wisconsin, U.S., Datiri and Lowery (1991a;b) observed that the wetting front moved faster in minimum tillage than in conventional tillage systems. Soils in minimum tillage systems had conductive pores which saturated the soil profile much quicker than in conventional tillage systems. Consequently, saturated hydraulic conductivity was also higher in minimum tillage than in plow tillage systems.

Effects of minimum tillage and mulch tillage systems on water transmission properties were evaluated for several soils in western Nigeria (Lal, 1987). Lal (1992) observed that infiltration rate declined rapidly following deforestation. However, the rate of decline was much more rapid in conventional tillage than in no-tillage method of seedbed preparation. The data in Table 9 show that infiltration rate, soil water sorptivity and soil water transmissivity were all higher in no-tillage compared with plowed treatments. These beneficial effects of no-tillage were primarily related to

Table 9. Effect of no-tillage and conventional tillage methods on infiltration characteristics of soils in watershed cleared by manual methods

	No-tillage (Watershed 3)				Conventional tillage (Watershed 1)			
Characteristics	Pre-clearing	1979	1980	1981	Pre-clearing	1979	1980	1981
Accumulative infiltration (cm/2 hr)	320.0	-	-	-	179.0	-	-	-
Infiltration rate (cm/hr)	110.0	30.6	28.2	12.6	78.0	18.6	8.4	7.2
Transmissivity (cm/hr)	1.9	0.39	0.35	0.17	0.86	0.24	0.04	0.03
Sorptivity ($cm/hr^{1/2}$)	2.1	0.43	0.67	0.36	1.50	0.24	0.55	0.27

(From Lal, 1992.)

Table 10. Tillage effects on runoff and soil erosion on Alfisols

	Plow till				
Year	Rainfall[a] (mm)	Runoff (mm)	Erosion (mg ha^{-1})	Runoff (mm)	Erosion (mg ha^{-1})
1975	1089	246.1	6.6	81.3	1.9
1977	310	21.5	1.3	12.8	0.6
1979	841	225.1	0.1	21.5	5.5
1980	900	34.4	0.3	153.0	1.9

[a] Rainfall for which soil erosion was measured.
(From Lal, 1984.)

less intensity of surface seal and hard setting, more activity of earthworms and termites, and continuity of root channels maintained in no-tillage treatments.

(ii) Mechanical measures:Engineering techniques are used to increase surface detention capacity and increase time available for water to infiltrate into the soil. Ridge tillage systems, contour ridges, and tied ridges are commonly used in soils with low infiltration rate. These techniques retain surplus water on the surface until it infiltrates into the soil. These minimum tillage systems are effective measures to increase infiltration in soils of the arid and semi-arid ecoregions. In Pakistan, Abhul Majid et al. (1986) observed that deep chiselling and tied ridge systems conserved water and increased maize yield by 95% under rainfed conditions. Similarly, beneficial effects of tied ridges were observed in regions of sort growing periods in Ethiopia (Araya, 1989). Highest soil moisture under Ethiopian conditions was observed for tied ridges constructed at 80 cm. intervals in a black soil at Mekele and at 40 cm. intervals in a red sandy soil at Quihe. Beneficial effects of tied ridges on water infiltration for the Sudan Savannah region of Burkina Faso was reported by Hulugalle (1988; 1989). This system of decreasing runoff losses is extremely effective for these soils of low infiltration capacity.

(iii) Erosion control: Increase in infiltration rate by minimum tillage systems leads to erosion control due to decreases in runoff amount and its sediment carrying capacity. The beneficial effects of minimum tillage systems in erosion control have been widely recognized. (Osuji, 1986; Unger, 1990; Lal, 1991b), and need not be repeated here. Crop residue mulch, however, is an important factor affecting the effectiveness of minimum tillage systems in erosion control. The data in Table 10 by Lal (1992) from western Nigeria show the importance of such a system in erosion control.

VI. Increasing Soil-Water Storage

In accord with the increase in infiltration and reduction in runoff losses, minimum tillage systems also increase soil water storage. There are several mechanisms responsible for increase in soil water storage (Figure 6). In addition to hydrological effects of altering components of soil-water balance, principal benefits of soil-water storage may be related to positive alterations in soil structure, e.g., changes in pore size distribution, moisture retention characteristics, and effective rooting depths.

There are several examples of higher soil water storage in the root zone in minimum tillage systems compared with plow-based methods of seedbed preparation (Blevins et al., 1971; 1977). Similar observations are reported from the Sudan Savanna region of Burkina Faso (Hulugalle, 1987) and Pakistan (Khokher and Nizami, 1987). In Darling Downs, Australia, Marley and Litter (1990) observed that water use efficiency (% of the rainfall stored in the rootzone) was 14% for stubble burning and conventional cultivation, 19.8% for stubble retained and conventional tillage, and 25.3 % for stubble retained and no-tillage.

(i) Alterations in p^F curves: Tillage-induced alterations in soil structure lead to favorable effects on soil moisture retention characteristics. An example of the effects of a no-tillage system on moisture retention characteristics of an Alfisol in western Nigeria is shown by the data in Table 11. The data on gravimetric moisture retention show only modest increase. However, soil bulk density of the surface 0-5 cm. layer was also more for no-tillage compared with plowed treatments. Consequently, the available moisture retention capacity on volumetric basis was measurably higher in no-tillage compared with plowed treatments. Higher moisture retention coupled with lower maximum soil temperature enhance water use efficiency and decrease drought susceptibility of these soils.

(ii) Changes in effective rooting depth: Biochannels, the bridging effect of roots connecting surface and subsoil for soil moisture use, and placement of fertilizers in subsoil through appropriate minimum tillage techniques can increase the effective rooting depth for exploitation of soil moisture reserves. On a Vertisol in Australia, Dalal (1989) observed that available water and NO_3-N were lower in the surface soil but more in the subsoil horizons (0.6-1.2 m depth) under no-tillage compared with conventional tillage systems. In another study, Dalal (1992) observed that water use efficiency and N recovery were higher in soil in no-tillage stubble retention treatment. In Morocco, Mohammed et al. (1991) observed that chisel tillage improved infiltration and soil water reserves to 20 cm. depth.

The data in Table 12 from a sandy soil in India show that minimum tillage and reduced tillage systems increased forage yield of maize especially when used in combination with crop residue mulch. The highest water use efficiency of 109.4 kg/ha/cm was obtained for the reduced tillage system with residue mulch. An example of the positive effects of mulch

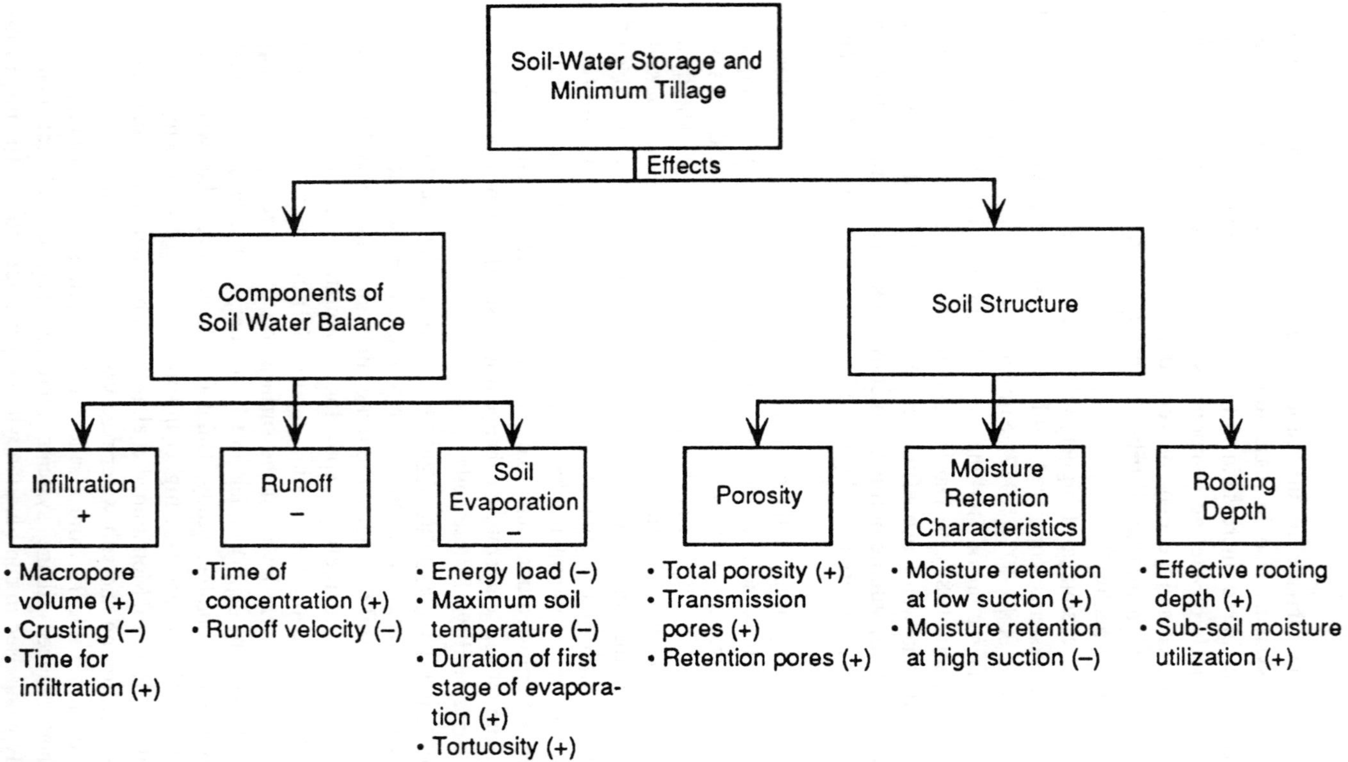

Figure 6. Processes affecting soil-water storage in the subsoil with minimum tillage, (+) refer to increase and (-) to decrease in capacity or intensity of a given parameter.

Table 11. Effects of no-tillage systems on soil moisture retention of a 0-5 cm depth in 1980, 3 years after initiating the experiment on Alfisols in Western Nigeria

Soil property	Tillage method	Soil moisture potential (M Pa) 0	0.003	0.006	0.01	0.03	0.05	1.5
Moisture retention (g /g, %)	Plow tillage	46.6 +/- 2.2	32.0 +/- 2.7	21.6 +/- 2.0	20.0 +/- 2.8	16.4 +/- 2.8	12.3 +/- 2.6	9.3 +/- 0.9
	No tillage	46.5 +/- 8.2	33.3 +/- 6.9	21.4 +/- 5.0	20.5 +/- 6.4	16.8 +/- 5.0	13.2 +/- 2.8	9.7 +/- 1.7

Table 12. Tillage and mulching effects on water use efficiency of forage maize in sandy soil in Frizabad, India

Tillage method	Mulch rate	Dry matter yield ($Mg\ ha^{-1}$)	Water use efficiency ($kg\ ha^{-1}\ cm^{-1}$)
Minimum tillage	0	4.62a	103.4
Reduced tillage	0	5.33b	101.7
Conventional tillage	0	5.60b	90.2
Minimum tillage	5	5.65b	109.1
Reduced tillage	5	6.87c	109.4
Conventional tillage	5	7.36c	103.7

(From Sharma, 1991.)

and no-tillage system on available soil-water reserves, increase in soil water storage over fallow, and grain yield of sorghum for an Australian Vertisol is shown by the data in Table 13. Thornburn (1992) observed that the highest grain yield, maximum available water stored to 1.8 m depth, and the greatest increase in soil water storage over the conventional fallow system occurred in no-tillage systems with crop residue mulch.

There is a strong theoretical basis and a convincing practical evidence that support the hypothesis that in ecoregions where soil moisture scarcity is a principal constraint, crop production and water use efficiency can be improved by a judicious and discriminate use of minimum tillage systems in combination with surface application of crop residue mulch.

Table 13. Tillage and residue management effects on soil moisture storage and sorghum yield on an Australian Vertisol

Tillage/residue treatment	Grain yield (Mg ha^{-1})	Available soil water to 1.8 m (mm)	Increase in soil water overflow (mm)
No-tillage +R	3.1	149	73
No-tillage -R	2.4	119	55
Blade plow +R	3.1	140	68
Blade plow -R	3.0	138	66
Disc plow +R	3.1	147	64
Disc plow -R	2.8	136	62

(From Thorburn, 1992.)

VII. Minimum Tillage and Rice

Rice is a semi-aquatic plant, and thrives better in almost saturated or inundated soil conditions. Tillage methods that enhance ponding are the most suited for this crop. Minimum tillage systems of seedbed preparation have several pertinent applications in rice-based cropping systems. Examples of some of these applications include the following:

(i) Rice-upland cropping system: Minimum tillage system for rice production is useful for the following upland crops, e.g., wheat (*Hordeum vulgare*), soybean (*Glycine max*), etc. Puddling and intensive tillage methods destroy soil structure, and have adverse effects on growth and yield of the following upland crop. Experiments conducted in the Taihu Lake region of China by Zhao (1989) shoed that a minimum tillage system was more suitable for paddy soils under rice (*Oryza sativa*)-wheat cropping system than conventional tillage.

(ii) Heavy-textured soils: Puddling or wet tillage to deliberately destroy soil structure for increasing soil-water retention and decreasing permeability is not as necessary for clayey soils of low permeability as for light-textured soils of high permeability. Experiments conducted in the Philippines showed significant differences among tillage methods, in daily water use (e.g., percolation, seepage and evaporation). Sharma, et al. (1988) observed less differences in water use among tillage treatments in clayey soils compared with clay loam soils. The daily water use (mm/day) in clayey soil and clay loam soil respectively was 37 mm and 393 mm for no-tillage dry sown method, 35 mm and 501 mm for no-tillage transplanted method, 34 mm and 263 mm for minimum tillage, and 30 mm and 161 mm for puddling method of rice cultivation. In another experiment from the Philippines, Mambani et al. (1990) concluded that there were no significant effects of tillage methods on soil properties and crop yields of clayey soils. In contrast, tillage methods had significant effects on water use and crop yields in clay loam soil. Weed infestation is, however, often higher with minimum than conventional tillage systems.

In accord with data from the Philippines, results of a long-term experiment conducted in western Nigeria showed differences in tillage effects on soil properties and rice yield among soils of contrasting texture. No-tillage and minimum tillage systems produced significantly low yields in a coarse-textured soil, but satisfactory yields equivalent to puddling were obtained in heavy-textured soils. Similar observations were made by Rodriguez and Lal (1985). Minimum tillage systems can, therefore, be conveniently adopted for rice-based cropping systems on relatively heavy-textured soils of the tropics and sub-tropics.

VIII. Minimum Tillage and Agricultural Sustainability

In view of the increasing population pressure and decreasing resource base, the concept of agricultural sustainability implies an increasing trend in per capita productivity to meet the present needs while enhancing productive capacity of the natural resources. Minimum tillage systems can play an important role in achieving agricultural sustainability (Lal, 1991a). There are several examples of higher production and more efficient use of resources by minimum tillage compared with conventional tillage systems. In some specific soil-related constraints to crop and forage production, minimum tillage systems can be used advantageously to enhance productivity, obtain similar yields with lower input, or similar yields with less risks of oil and environmental degradation. On a Vertisol in Puerto Rico, for example, Lugo-Mercado (1988) observed similar yields of several vegetable crops among minimum tillage and conventional tillage systems of seedbed preparation. One of the major and common limitations of no-tillage system in relation to productivity is the weed control. Heavy weed infestation is a common problem in most minimum tillage systems.

Relevant examples of tillage effects on productivity of the contrasting ecoregions are shown by the data in Tables 14 and 15, respectively. For an Ultisol in humid regions of eastern Nigeria, Ohiri and Ezumah (1990) observed that cassava grown by no-tillage and minimum tillage systems outyielded the conventional system by 40 and 23%, respectively (Table 14). In fact, the stem yield (used for replanting the next season crop) was significantly more in no-tillage compared with conventional tillage system. In comparison with initial carbon content of 2.4%, soil organic carbon declined in all treatments. However, the decline was most severe in conventional tillage compared with no-tillage and minimum tillage systems. In semi-arid region of northern Nigeria, Adeoye and Mohamed-Saleem (1990) evaluated effects of tillage methods on maize grain yield and dry matter yield of stylo (*Stylsanthes hamata*). The data in Table 15 show significantly more maize yield and stylo dry matter yield in subsoiling compared with disc harrowing treatment. Subsoiling lowered bulk density by 0.04-0.09 Mg/m^3 for 0-10 cm. layer, and by 0.19-0.32 Mg/m^3 for 10-30 cm.

Table 14. Tillage method effects on cassava tuber yield and soil organic carbon content of an Ultisol in Nigeria

Tillage method	Tuber yield ($Mg\ ha^{-1}$)	Stem yield ($Mg\ ha^{-1}$)	Soil organic carbon (%)
No-tillage	13.0	27.8	1.84
Minimum tillage	12.1	26.1	1.87
Conventional tillage	12.0	21.2	1.64
LSD (0.05)	NS	6.3	0.15

Initial soil organic carbon content = 2.4%.
(From Ohiri and Ezumah, 1990.)

Table 15. Tillage method effects on maize grain yield and dry matter yield of Stylo on a degraded soil in northern Nigeria

Tillage method	Maize grain yield ($kg\ ha^{-1}$)	Stylo dry matter yield ($kg\ ha^{-1}$)
Sub-soiling	4926	6865
Disc harrowing	3962	5480
Ridging	4210	5037
LSD (0.05)	838	843

(From Adeoye and Mohamed-Saleem, 1990.)

layer. Maize grain yield increased by 6% by ridge tillage and by 24% by subsoiling.

Lal (1992) conducted a 10-year watershed management experiment in western Nigeria and showed a drastic reduction in maize grain yield with time. The rate of yield decline was 0.2/Mg/yr. This rapid decline in yield was attributed to degradation of soil physical, chemical and biological properties. Although soil degradation was relatively less and maize grain yield was relatively more in minimum tillage compared with conventional tillage system, the degradation with mechanized systems of farming was unavoidable on these structurally fragile soils in harsh tropical environments. In addition to the minimum tillage, there is a need to develop ecologically compatible land use systems and less intensive farming systems. Some environments cannot support intensive row cropping despite the use of minimum tillage systems.

IX. Ecological Limits of Minimum Tillage Systems

Minimum tillage systems are not suited for all soils, crops, and ecoregions (Lal, 1985). In unsuitable environments, minimum tillage systems can lead to soil compaction, weed infestation, and reduced yields. Most suitable conditions for use of minimum tillage systems are:

(i) Soil conditions: Minimum tillage systems are most suited for medium textured, well-drained, and structurally active soils. Soil conditions for good crop growth and sustained production are not optimum in poorly drained soils, heavy-textured soils with massive structure, and soils prone to compaction. Minimum tillage systems can also be adopted for some soils with other ameliorative measures, e.g., surface and internal drainage, land forming, subsoiling to eliminate plow pans, etc.

(ii) Cropping systems: Minimum tillage systems are generally suited for grain crops in a monocropping system. Although minimum tillage can be adopted for mixed cropping systems, there may be a problem of selective herbicides. There may also be problems in using minimum tillage techniques for production of some root crops, e.g., yam. Root crops can be successfully grown with raised bed types of minimum tillage. Effectiveness of minimum tillage techniques can be greatly enhanced when used in association with residue mulch. Crops and cropping systems that produce enough residue for use as a mulch for the succeeding crop are readily adapted to minimum tillage techniques, e.g., cereals-legumes, cereals-cereals, cereal-root crops, etc.

(iii) Ecoregions: Appropriate soil type and suitable cropping systems for profitably using minimum tillage systems are found in humid-sub-humid and semi-humid ecoregions. Therefore, minimum tillage systems are less suited for arid and semi-arid ecoregions. Adoption of minimum tillage techniques for arid regions require special ameliorative measures, e.g., tied ridges, rough seedbed, etc.

Minimum tillage systems are suited to alleviate specific soil-related constraints to intensive cultivations. These constraints include: (i) accelerated soil erosion by water and/or wind, (ii) drought stress and low water use efficiency, (iii) supra-optimal soil temperatures, (iv) soils prone to crusting and hard-setting, and (vi) soils prone to the formation of a plow pan.

Several attempts have been made to develop soil guides and environmental indices for assessing soil's suitability for adopting an appropriate minimum tillage system (Lal, 1985). Applicability of minimum tillage systems require development of specific packages of cultural practices, e.g., cropping systems, residue management, type and method of fertilizer application, need control measures, water management (drainage/irrigation) techniques, etc. Most of these practices are soil and site-specific and require local adaptive on-farm research.

X. Summary and Conclusions

Minimum tillage system is an ecological approach to resource management. There is a wide range of minimum tillage systems designed to alleviate specific soil-related constraints to crop production. An important application of minimum tillage system is in soils and ecoregions prone to accelerated erosion, and regions prone to drought stress. Despite potential advantages, there are several issues which require research attention:

(i) Tilth index: Soil tilth is still a major unknown that is difficult to quantify. Several attempts have been made to develop such an index. For Vertisols, McGarry (1988) proposed using shrinkage curve indices in quantification of the effects of minimum tillage systems on soil properties. Colvin, et al. (1990) and Karlen, et al. (1990) have developed a tilth index. There is a need to develop a quantitative and routinely usable index of tilth assessment in relation to soil tillage.

(ii) Soil guide for tillage methods: There is a need to develop a soil guide for choosing appropriate tillage techniques for alleviating soil-related constraints to crop production in different ecoregions. Soil guide for tillage requirements should be developed on the basis of a broad-based data bank on crop performance in different tillage methods with regards to tillage-induced alterations in soil properties, erosion, risks, water and nutrient use and losses, and environmental concerns.

(iii) Evolution of soil properties: Long-term changes in soil properties under different systems of minimum tillage are not known for most soils, especially those in the tropics. Understanding dynamics of soil physical, chemical, and biological properties is crucial to developing systems of soil tillage that minimize degradation and enhance and sustain productivity. Evolutionary changes in two soil properties have far-reaching economic and ecologic impact, e.g., soil organic matter and soil structure, and must be studied over a long time.

(iv) Soil compaction: Densification of the surface or subsoil can be a problem under minimum tillage systems. Soil compaction is likely to be a severe constraint with heavy-textured soils, poor drainage and soil wetness, intensive row crop farming, and mechanized cultivation. This problem needs to be addressed by a multi-disciplinary team with interests in hydrology, soil science, machinery, crop rotations, and farming systems.

(v) Tillage and non-crop period: Soil degradation, and excessive loss of water and nutrients form the root zone, can occur during the non-crop or fallow periods. Type of minimum tillage system for erosion control, moisture conservation, and weed control need to be developed especially for hard-setting soils of the semi-arid and semi-humid ecoregions.

(vi) Zonal tillage: Special attention is needed for minimum tillage techniques for row-interrow management. The row area should be managed differently than inter-row area for sustainable use of soil resources. Minimum tillage needs for row-interim management may be different for different soils and cropping systems, e.g., wheat traffic zone, furrow in raised beds, etc.

(vii) Tillage and global change: Soil tillage plays an important role in the carbon dynamics in the plow zone. Little is known regarding the influence of minimum tillage systems on flux of radiatively-active gases from soil to the atmosphere. What tillage systems are appropriate for carbon sequestration in soil, and for minimization of gaseous flux?

(viii) Ecological and holistic approach: Tillage plays an important role in successful and sustainable management of agricultural ecosystems. Therefore, minimum tillage systems should be developed and fine-tuned with due consideration to ecological principles, especially the application of basic principles from natural ecosystem to agricultural ecosystems. Successful minimum tillage systems are those that are developed with incorporation of ecological principles.

References

Abhul Majid, M. Shafiq, and M. Iqbal. 1986. Deep tillage and sowing techniques in maize production under high rainfed conditions. *Pakistan J. Agric. Res.* 7: 181-185.

Adeoye, K.B., and M.A. Mohamed-Saleem. 1990. Comparison of effects of some tillage methods on soil physical properties and yield of maize and stylo in a degraded ferruginous tropical soil. *Soil and Tillage Res.* 18: 63-72.

Agarwal, R.P. 1991. Water and nutrient management in sandy soils by compaction. *Soil & Tillage Res.* 19: 121-130.

Araya, W.A. 1989. Effect of ridge spacing on wheat yield in areas with short growing periods. *Ethiopian J. Agric. Sci.* 1: 85-87.

Barnasor, P.C. and S.K. De Datta. 1988. Long-term effects of tillage, cultivar and herbicide on weed shift and control in broadcast-seeded flooded rice. *Soil & Tillage Res.* 12: 197-212.

Blevins, R.L., D. Cook, S.H. Phillips, and R. E. Phillips. 1971. Influence of no-tillage on soil moisture. *Agron. J.* 63: 593-596.

Blevins, R.L., G.W. Thomas, and P.L. Cornelius. 1977. Influence of no-tillage and N fertilization on certain soil properties after 5 years of continuous corn. *Agron. J.* 69: 383-386.

Brooke, H.D., D.R. Coventry, T.G. Reeves, and D.K. Jarvis. 1989. Liming and deep ripping responses for a range of field crops. *Plant and Soil* 115: 1-6.

Centurion, J.F. 1987. Effect of different tillage methods on the physical properties of a cerrado soil and on maize yield. *Cientifica* 15: 1-8.

Chan, K.Y. and J.A. Mead. 1988. Water movement and macroporosity of an Australian alfisol under different tillage and pasture conditions. *Soil & Tillage Res.* 14: 301-310.

Chan, K.Y. and J.A. Mead. 1988. Surface physical properties of a sandy loam soil under different tillage practices. *Aust. J. Soil Res.* 26: 549-559.

Charreau, C. 1977. Controversial points in dryland farming practices in semi-arid West Africa. p. 313-360. In: G.H. Cannell (ed.) "Symposium Proc. Rainfed Agric. in Semi-arid Regions", Univ. California, Riverside.

Chopart, J.L. and R. Nicou. 1989. Twenty years of continous cropping with and without cultivators in Senegal. *Agron Tropicale* 44: 269-281.

Chopart, J.L., J.M. Kalms, J. Marquett and R. Nicou. 1981. *Comparison de differentes techniques de travail du sol en trois ecologies de l'Afrique de l'Ouest.* IRAT, Montpollier, France. 65 pp.

Colvin, T.S., A.Q. Mughal, K.K. Singh and S. Thangavadivelu. 1990. A tilth index for tillage management. p. 295-301. In: V.M. Salokhe, and S.G. Ilangontileke (eds.) Proc. Intl. Agric. Eng. Conference and Exhibition. 3-6 Dec. 1990, Bangkok, Thailand.

Cornish, P.S. and J.E. Pratley (eds.). 1987. *Tillage: New Directions in Australian Agriculture.* CSIRO, Canberra, Australia.

Costa, P., J.L. Chopart, J.L. Janeau, and C. Valentin. 1989. Runoff measurement on a gravelly soil after six years of continuous cultivation with or without ploughing in the Ivory Coast. *Agron. Trpicale* 44: 255-262.

Dalal, R.C. 1989. Long-term effects of no-tillage, crop residue, and nitrogen application on properties of a Vertisol. *Soil Sci. Soc. Am. J.* 53: 1511-1515.

Dalal, R.C. 1992. Long-term trends in total nitrogen of a vertisol subjected to zero-tillage, nitrogen application and stubble retention. *Aust. J. Soil Res.* 30: 223-231.

Datiri, B.C. and B. Lowery. 1991a. Tillage effects on rain-generated wetting front migration through a Griswold silt loam soil. *Soil & Tillage Res.* 21: 243-256.

Datiri, B.C. and B. Lowery. 1991b. Effects of conservation tillage on hydraulic properties of a Griswold silt loam soil. *Soil & Tillage Res.* 21: 257-271.

Dongale, J.H. 1987. Effects of cropping system and tillage on infiltration characteristics of medium black soil of Konkan. *J. Maharashtra Agric. Univ.* 12: 9-11.

Gebhardt, M.R., A.K. Creek, C.L. Webber, and J.M. Gregory. 1986. Soybean culture effects on compaction of a clay pan soil. *Trans. ASAE* 29: 416-420.

Hadda, M.S. and H.S. Sur. 1989. Effect of mulch rates on runoff, sediment and nutrient losses from a sandy soil under untilled and tilled conditions. *J. Res. Pb. Agric. Univ.* 26: 37-46.

Holloway, R.R. and A.R. Dexter. 1991. Tillage and compaction effects on soil properties, root growth and yield of wheat during drought in a semi-arid environment. *Soil Technology* 4: 233-253.

Hulugalle, N.R. 1987. Effect of tied ridges on soil water content, evapotranspiration, root growth and yield of cowpeas in the Sadan Savanna of Burkina Faso. *Field Crops Res.* 17: 219-228.

Hulugalle, N.R. 1988. Intercropping millet and bamabara groundnut on tied ridges in the Sadan Savannah of Burkina Faso. *Arid Soil Res. & Rehabilitation* 2: 97-109.

Hulugalle, N.R. 1989. Effect of tied ridges and undersown *Stylosanthes hamata* on soil properties and growth of maize in the Sudan Savannah of Burkina Faso. *Agric. Escosystem and Environment* 25: 39-51.

Karlen, D.L., D.C. Erbach, T.C., Kasper, T.S. Colvin, E.C. Berry and D.R. Timmons. 1990. Soil tilth: a review of past perceptions and future needs. *Soil Sci. Soc. Am. J.* 54: 153-160.

Khan, B.R. and S.H. Raza. 1989. Effect of tillage and water harvesting techniques on the yield of cereals and food legume under rainfed conditions. *Sarhad J. Agric.* 5: 331-335.

Khokhar, M.A. and M.I. Nizami. 1987. Effect of deep tillage as moisture conservation technique on yield of rainfed wheat. *Pakistan J. Agric.* Res. 8: 12-16.

Lal, R. 1984. Mechanized tillage systems effects on soil erosion from an Alfisol in watersheds cropped to maize. *Soil Tillage Res.* 4: 349-360.

Lal, R. 1985. A soil suitability guide for different tillage systems in the tropics. *Soil & Tillage Res.* 5: 179-196.

Lal, R. 1986. Effects of 6 years of continuous no-till or puddling systems on soil properties and rice yield of a loamy soil. *Soil & Tillage Res.* 8: 181-200.

Lal, R. 1987. *Tropical Ecology and Physical Edaphology.* J. Wiley & Sons, U.K.

Lal, R. 1988. Residue management, nitrogen, and tillage effects on grain yield of paddy rice for a sandy hydromorphic soil in western Nigeria. p. 727-731. In: Proc. 11th Conf. ISTRO, Edinburgh, U.K.

Lal, R. 1990. Ridge tillage. *Soil & Tillage Res.* 18: 107-111.

Lal, R. 1991a. Tillage and agricultural systainability. *Soil & Tillage Res.* 20: 133-146.

Lal, R. 1991b. *Soil Erosion in the Tropics: Principles and Management.* McGraw Hill, New York.

Lal, R. 1992. *Tropical agricultural hydrology and sustainability of agricultural systems: a ten year watershed management project in southwestern Nigeria.* The Ohio State University, IITA Bulletin, Columbus, Ohio. 303 pp.

Lal, R. 1993. Global soil erosion by water and carbon dynamics. Proc. International Symposium *Soil Processes and Management Systems: Greenhouse Gas Emissions and Carbon Sequestration*, 5-9 April, 1993, Columbus, Ohio.

Lal, R., D.J. Eckert, N.R. Fausey, and W.M. Edwards. 1990. Conservation tillage in sustainable agriculture. p. 203-223. In C.A. Edwards et al. (eds.) *Sustainable Agricultural Systems*, Soil Water Cons. Soc. Ankeny, IA.

Lugo-Mercado, H.M., J. Badillo-Feliciano, E. Orengo, C. Ortiz and F.H. Beinroth. 1988. Yield response of vegetables to tillage techniques on a vertisol of the semi arid region of Puerto Rico. *J. Agric. Univ. Puerto Rico* 72: 119-125.

Lyle, W.M. and J.P. Bordovsky. 1981. Low energy precision application (LEPA) irrigation system. *Trans. ASAE* 24: 1241-1245.

Mambani, B., S.K. De Datta and C. A. Redulla. 1990. Soil physical behavior and crop responses to tillage in lowland rice soils of varying clay content. *Plant and Soil* 126: 227-235.

Marley, J.M. and J.W. Litter. 1990. Winter cereal production on the Darling Downs: a comparison of reduced tillage practices. *Aust. J. Expl. Agric.* 30: 83-93.

McGarry, D. 1988. Quantification of the effects of zero and mechanical tillage on a vertisol by using shrinkage curve indices. *Aust. J. Soil Res.* 26: 537-542.

Mead, J.A. and K.Y. Chan. 1988. Effect of deep tillage and seedbed preparation on the growth and yield of wheat on a hard-setting soil. *Aust. J. Expl. Agric.* 28: 491-498.

Mohammed, T., C.W. Gay and R. E. Banner. 1991. The effects of chiseling and S fertilzation on infiltration, soil moisture and vegetation parameters in the Ait Rbaa Perimeter of Central Morocco. *Arid Soil Res & Rehabilitation* 5: 271-288.

Moran, C.J., A.J. Koppi, B.W. Murphy, and A. B. McBratney. 1988. Comparison of the macropore structure of a sandy loam surface soil horizon subjected to two tillage treatments. *Soil Use & Management* 6: 96-102.

Nicou, R. 1977. *Le teravail du sol dans les terres exondees du Senegal: Motivations, containtes.* ISRA/CNRA, Bambay, Senegal, Mimeo. 50 pp.

Nimmo, J.R. and C.A. Katherine. 1988. Hydraulic conductivity of a sandy soil at low water content after compaction by various methods. *Soil Sci. Soc. Am. J.* 52: 303-310.

Ohiri, A.C. and H.C. Ezumah. 1990. Tillage effects on cassava production and some soil properties. *Soil & Tillage Res.* 17: 221-229.

Oni, K.C. 1991. Traffic compaction and tillage effects on the performance of maize in sandy loam soil of Nigeria. Agric. *Mechanization in Asia, Africa and Latin America.* 22: 27-31.

Onwualu, A.P. and U.G.N. Anazoda. 1989. Soil compaction effects on maize production under various tillage methods in a derived savanna zone of *Nigeria. Soil & Tillage Res.* 14: 99-114.

Osuji, G.E. 1986. Zero-tillage for erosion control and maize production in the humid tropics. *J. Envir. Mgmt.* 23: 193-201.

Oussible, M. and R.K. Crookston. 1987. Effect of sub-soiling a compacted clay loam soil on growth, yield and yield components of wheat. *Agron.J.* 79: 882-886.

Prove, B.G., R.J. Loch, J.L. Foley, V.J. Anderson, and D.R. Younger. 1990. Improvements in aggretation and infiltration characteristics of a Krasnozem under maize with direct drill and stubble retention. *Aust. J. Soil Res.* 28: 577-590.

Raghavan, G.S.V., P. Alvo and E. McKyes. 1990. Soil compaction in agricuture. *Adv. Soil Sci.* 11: 1-36.

Rahman, S.M. and A. Islam. 1989. Effects of tillage depth on infiltration characteristics of two Bangladesh soils having ploughpans. *Soil & Tillage Res.* 13: 407-412.

Resource Conservation Glossary. 1982. Soil Conservation Society of America, Ankeny, Iowa.

Ritchie, J.D. 1983. p. 8:1-6. In: Source book for farm energy alternatives. McGraw Hill Book Co. NY.

Rodriguez, M.A. and R. Lal. 1985. Growth and yield of paddy rice as affected by tillage and nitrogen levels. *Soil & Tillage Res.* 6: 163-178.

Roth, C.H., B. Meyer, H.G. Frede, and R. Derpsch. 1988. Effect of mulch rates and tillage systems on infiltrability and other soil physical properties of an Oxisol in Parana, *Brazil. Soil & Tillage Res.* 11: 81-91.

Roth, C.H., B. Meyer, H.G. Frede and R. Derpsch. 1986. The effect of different soybean tillage systems on infiltrability and erosion susceptibility of an Oxisol in Parana. *Brazil. J. Agronomy & Crop Sci.* 157: 217-226.

Rusanov, V.A. 1991. Effects of wheel and track traffic on the soil and on crop growth and yield. *Soil & Tillage Res.* 19: 131-143.

Sharma, B.R. 1991. Effect of different tillage practices, mulch and nitrogen on soil properties, growth and yield of fodder maize. *Soil & Tillage Res.* 19: 55-66.

Sharma, P.K., S.K. De Datta and C.A. Redulla. 1988. Tillage effects on soil physical properties and wetland rice yield. *Agron. J.* 80: 34-39.

Sidiras, N. and C.H. Roth. 1987. Infiltration measurements with double-ring infiltrameters and a rainfall simulator under different surface conditions on an oxisol. *Soil & Tillage Res.* 9: 161-168.

Sommer, C. and M. Zach. 1992. Managing traffic induced oxisol compaction by using conservation tillage. *Soil & Tillage Res.* 24: 319-326.

Steinert, P., D. Werner, and J. Lhotsky. 1990. The effects of rooting on the structure of compacted soil zones after slit ploughing. *Scientia Agricultural Bohemosloveca* 22: 85-91.

Stout, B.A. 1984. Energy use and management in agriculture. Breton Publishers, North Scituate, MN.

Thorburn, P.J. 1992. Structural and hydrological changes in a vertisol under different fallow management techniques. *Soil & Tillage Res.* 23: 341-359.

Unger, P.W. 1990. Conservation Tillage Systems. *Adv. Soil Sci.* 13: 27-68.

Unger, P.W. and J.T. Musick. 1990. Ridge tillage for managing irrigation water on the U.S. Southern Great Plains. *Soil & Tillage Res.* 19: 267-282.

Vepraskas, M.J. 1988. Bulk density values diagnostic of restricted root growth in coarse-textured soils. *Soil Sci. Soc. Am.J.* 52:1117-1121.

Watson R. T., L. G. Meira Filho, E. Sanhueza and T. Janetos. 1992. Sources and sinks. In: J.T. Houghton, B.A. Callander and S.K. Varney (eds.) Climate Change. Cambridge Univ. Press, U.K.

Zhao, C.Z. 1989. Physical properties of paddy soils in relation to minimum and zero-tillage methods in the Taihu Lake region. *Acta Pedologica Sinica* 26: 101-108.

Improving and Managing Subsoil Structure for High Productivity from Temperate Crops on Beds

K.A. Olsson, B. Cockroft, and P. Rengasamy

I. Introduction

Dense, slowly-permeable subsoils around the world restrict the productive potential of soils and make the management of otherwise fertile, well-structured topsoils difficult. Scientists and farmers alike for long have sought to improve the structure of such subsoils by deep tillage, ameliorants, soil organisms, and ameliorative crops. They seek higher crop production through faster infiltration, deeper penetration of water, increased water availability, increased drainage, leaching of salts and toxins, release of nutrients, lower strength and deeper rooting. Soil profile modification, however, has not consistently increased crop

1-56670-020-5/95/$0.00+.50

yields in spite of many decades of effort (Wells, 1956; Eck and Unger, 1985). Benefits are usually transient and variable (Unger et al., 1981). This in large part stems from not understanding how to break up the soil to the best particle arrangement nor, importantly, how to stabilize it. A quantitative approach to profile modification is needed, based on specific design criteria. Research must integrate the soil-based disciplines of physics, chemistry, mechanics and biology and combine these with agronomic innovation (cf. Sharpley et al., 1992). Hence, this paper aims to develop simple practical guidelines for on-field management of different cropping systems.

In Australia, the yields of many crops, even when irrigated, are less than half the potential calculated from solar radiation and water supply (Cockroft and Martin, 1981; French and Schultz, 1984). The highest yields come from the soils which support the largest root systems. Yields close to potential have been achieved, but only on soils where physical and nutritional factors are ideal for roots in both the surface soil and the subsoil. Such soils are rare worldwide (for example, some soils developed on volcanic ash) and so our research for higher yields on most soils must include subsoil as well as surface soil improvement, and be directed at providing the root's needs for growth and functioning.

Soil scientists and farmers have achieved major increases in crop yield by improving soil physical properties and nutrition, but these improvements are mainly to the topsoil. While criteria for favourable structure of topsoils are well documented (Shaw, 1952; Greenland, 1981; Tisdall and Adem, 1988; Mullins et al., 1990), these are often difficult to achieve in practice without structural modification of the underlying subsoil. On shallow soils, improvement in physical properties of the topsoil allows only a limited improvement in water and nutrient supply to the crop due to the small volume of soil available for root exploitation.

Thus, in order to achieve and sustain high productivity from crops on such soils, we need to develop and stabilize the structure of clay subsoils. This requires an understanding of processes of structure formation and stabilization at all levels in the soil structural hierarchy, from micro- to macrostructure (Dexter, 1988).

This paper outlines an approach to improving structure in subsoils to enable sustained, high productivity from both irrigated and rainfed crops. The approach aims to promote root growth and function in the subsoil. We write from an Australian perspective based on experience with irrigated alfisols in a semi-arid environment. In Section II, we discuss the physical limitations to the growth and functioning of roots imposed by clay subsoils. In Section III, we suggest design criteria for subsoil structure and suggest methods for improving porosity, reducing strength and stabilizing these in modified clay subsoils. In Section IV, we discuss the requirements for soil management systems for crops where the farmer controls traffic, uses minimum tillage, and forms beds. He thus avoids stresses due to traffic compaction, cultivation and slaking by water.

II. Limitations on Yield Imposed by Poorly Structured Subsoils

Poorly structured subsoils constrain crop productivity firstly by restricting the growth and functioning of roots and secondly by restricting gas exchange and the storage and transport of water (and nutrients) through low porosity of the soil. We are concerned here mainly with soil physical constraints.

As well as taking up water and nutrients, plant roots also control the activity of shoots and physiological processes and act as storage organs (Gregory et al., 1987). The interdependence of roots and the aerial parts of the plant is crucial to yield so that impeded root growth, through poor soil structure, can decrease yield (Richards and Rowe, 1977; Passioura, 1991).

Roots also function in the development and maintenance of structure and in stabilizing soil aggregates (Tisdall and Oades, 1982) in association with fungi and in forming stable pore networks (see Section III).

A. Physical Restrictions on Root Growth Produced by Dense Subsoils

Given adequate levels of water, nutrients and temperature, roots need at least 10 mg O_2 L^{-1} soil in order to grow (Meyer and Barrs, 1991); this requires a minimum air-filled porosity of about 10% of the soil volume on drainage (Wesseling, 1974; Jayawardane and Meyer, 1985; Meyer et al., 1985). For roots to elongate, they need a soil strength lower than about 3 MPa, measured as penetrometer resistance (Taylor et al., 1966), but lower than 0.5 MPa for unrestricted growth (Cockroft et al., 1969). The aeration and strength requirements for root growth are discussed further in Section III.A.

Figure 1 illustrates, schematically, the restricted opportunity for roots to grow in a dense subsoil compared with a porous topsoil, where water, temperature and nutrients are assumed non-limiting for root growth.

In the dense subsoil, the range of matric suctions over which roots are able to grow, where not limited by low aeration or high strength, is small. The subsoil has to be dried to the suction B in order to meet the minimum air-filled porosity. However, soil strength increases with increasing matric suction and exceeds the limiting value for unrestricted root growth at low suction. The small range of suction represented by A-B thus represents a restricted opportunity for root growth as the soil dries. In contrast, the high porosity of the topsoil allows the aeration requirement to be met after only a small amount of drainage, to suction D. Soil strength, in this case, becomes limiting at the much higher suction, C, allowing roots to grow over a wider range of suction as well as for a longer time.

This concept is similar to the non-limiting water range (NLWR) for plant growth described by Letey (1985) and by Boone (1988).

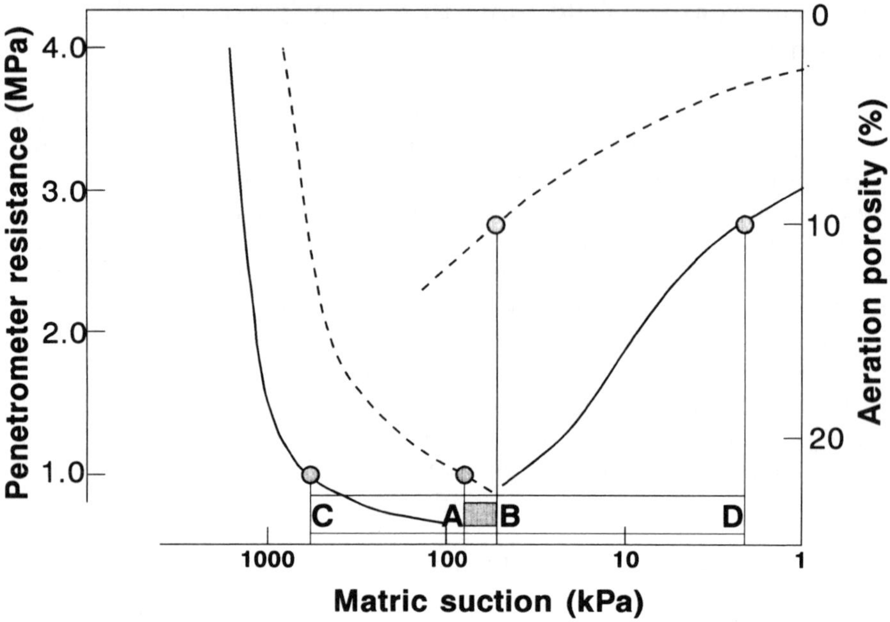

Figure 1. Schematic comparison of ranges of matric suction in contrasting soils where aeration and strength are non-limiting for root growth.

B. Development and Functioning of Root Systems

Poorly structured subsoils produce shallow, confined root systems. Most field studies show high concentrations of roots of various crops in the topsoil (Cockroft and Wallbrink, 1966; Greacen, 1977; Taylor and Olsson, 1987). But root concentrations are low in B horizons, hard pans, plough pans or layers with nutritional problems. Commonly, root concentrations diminish to less than 25% of those in the topsoil within the upper 500 mm of the soil. In flood-irrigated peach trees on an alfisol at Tatura, root concentrations decrease from 32 km m^{-3} soil in the topsoil and 8 km m^{-3} at 500 mm depth ; total root length beneath unit ground area is only 10.8 km m^{-2} to 1.2 m depth (Olsson and Rose, 1988). Root distribution is assumed to decrease exponentially with depth in models describing water and solute transport (e.g., Wagenet and Hutson, 1989; Ross et al., 1992).

Such shallow root systems on dense soils of low porosity and high strength cause crop management problems, limit the supply of water to the crop and limit plant size and vigor physiologically.

1. Crop Management Problems

The farmer must work with this shallow root system. His management becomes difficult and inflexible because the system is poorly buffered against drying, excessive wetness, high soil temperatures, high concentrations of fertilizers, possible toxicities and acidification. The farmer must, therefore, irrigate frequently, make special provision for drainage, keep the soil cool with mulches or quick-growing crop canopies and fertilize frequently. He must also monitor the topsoil frequently for toxins and pH. Further, the frequent irrigation and poor drainage cause the structure to deteriorate so that the soil becomes hard and permeability decreases. The farmer must resort to heavy dressings of organic manure or put the field into a pasture rotation; the former is costly, the latter is unprofitable.

2. Restrictions on Water Uptake by Roots in Subsoils

An extensive literature attests to positive linear relationships between water flow through the soil-plant system and yield in most temperate crops (Penman, 1971; contributors to Taylor et al., 1983). Therefore, any restriction on water supply to the crop imposed by shallow topsoils and poor subsoil structure will reduce yield.

As shown in Figure 1, the high mechanical strength and low aeration of dense clays restrict root growth and result in low root concentrations in the subsoil. Shallow wetting confines root systems to shallow depths due to the high mechanical strength of the dry subsoil. In addition, the low porosity of dense subsoils restricts the storage as well as rates of transport of water, gases and nutrients through the soil.

In discussing porosity in this paper it is convenient to use a number of pore classes; these are given in Table 1 together with the corresponding matric suctions.

Dense subsoils have few macropores and restrict both profile drainage and the depth of wetting following irrigation and rain. For example, the B horizons of many alfisols in southern Australia have fewer than 5% macropores by volume of soil (Olsson and Rose, 1978; Blackwell et al., 1990). The clay B horizon of Shepparton fine sandy loam (Skene and Poutsma, 1962) in northern Victoria, for example, has a saturated hydraulic conductivity of about 30 mm d^{-1} (Olsson and Rose, 1978). Soil horizons having saturated hydraulic conductivities lower than 60 mm d^{-1} are considered to be slowly permeable to water in the classifications cited by Marshall (1959).

Irrigation of such soils often produces near-saturation of the lower topsoil due to the low hydraulic conductivity of the B horizon. Protracted periods of near-saturation of both the topsoil and subsoil following irrigation and rain commonly result in low concentrations of oxygen that directly restrict both root growth and water uptake by roots (Cannell and Jackson, 1981). There is also an increased

Table 1. Classes of soil pores and corresponding matric suctions

Pore diameter (µm)	Category	Matric suction (kPa)
> 30	Macropores	< 10
30 - 0.2	Storage pores	10 - 1500
< 0.2	Residual pores	> 1500

risk of plant disease under these conditions from soil-borne pathogens (Bergman, 1959; Flett, 1986).

In dense subsoils, storage pores (Table 1) commonly comprise less than 10% of the soil volume or 10 mm dm^{-1} depth of soil. This contrasts with 20% (20 mm dm^{-1} depth) for many well-structured soils (Hall et al., 1977; Williams, 1983). An example is the vertisols, but compaction may lower the porosity even on these soils. The lack of storage pores also restricts the unsaturated hydraulic conductivity over the NLWR (Olsson and Rose, 1978) and, hence, rates of water and nutrient flow to the root surface.

As a result of the restricted volume of soil available for root proliferation, the low water storage and low rates of water movement through the subsoil, root systems cannot support transpiration rates higher than about 7 mm d^{-1} for more than a few days after irrigation in semi-arid environments. For many crops on shallow clay soils, transpiration rates may decrease to 50% of the potential after only 5-7 days or < 50 mm water transpired (Willatt and Olsson, 1982; Rab and Willatt, 1987). This is a major limitation to yield and necessitates frequent irrigation to avoid water stress in the crop. In deep soils, by contrast, extensive root systems tap a large reservoir of water more uniformly and sustain high transpiration rates over time (Veihmeyer and Hendrickson, 1938).

The dynamics of water supply are illustrated by a detailed analysis of water withdrawal patterns by peach trees under flood irrigation by Olsson and Rose (1988) on the Shepparton soil. As shown in Figure 2, transpiration decreases below the potential rate when the trees depend on water stored in the subsoil below 400 mm depth. In their study, the shallow topsoil below cultivation depth stored less than 20 mm available water. However, even where the whole topsoil is formed up into beds > 300 mm deep, as described by Cockroft and Tisdall (1978), the poor supply from the subsoil remains a limitation to continued water supply over time at the potential transpiration rate.

In the shallow topsoil, the high root concentrations and high soil conductances for water (calculated from the steady-state model of Gardner, 1960), allow the roots to fully exploit the limited water storage. In the subsoil, however, there are fewer roots than in the surface soil; rates of water uptake by roots are also lower when compared at similar matric suctions. While there is a close dependence of water uptake on root concentration over depth in the subsoil, rates of water uptake per unit root length are appreciably lower than in the topsoil at similar matric suctions. For example, the highest rate of water uptake per unit root

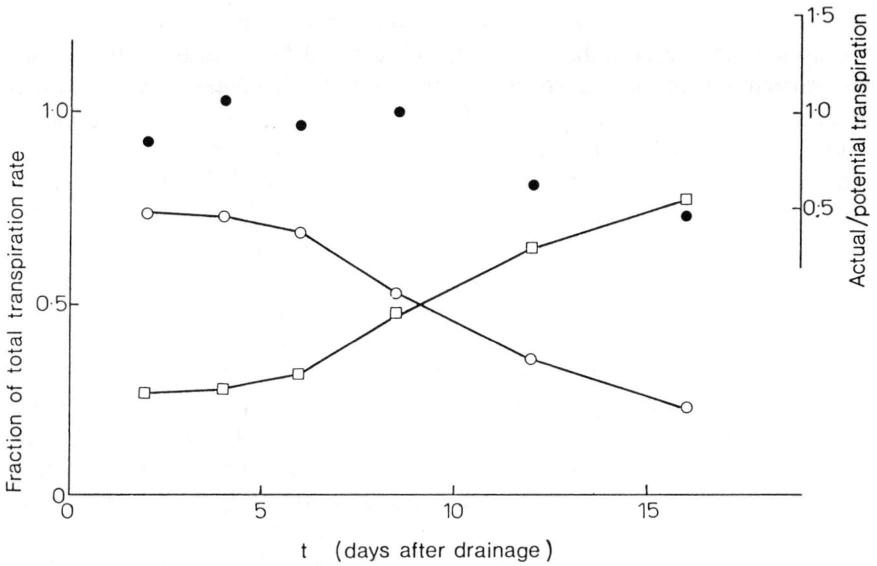

Figure 2. Fraction of total transpiration supplied from depth groupings in the soil profile during a 17-day drying cycle after irrigation in peach trees. Depth interval 0 - 0.4 m (open circles); 0.4 - 1.2 m (squares). Filled circles show actual/potential transpiration rates for each period.
(From Olsson and Rose, 1988.)

length for roots in the B horizon is only about half that for the topsoil when compared at the same matric suction. That is, the lower rate of water uptake by roots in the B horizon is not fully explained by the lower root concentrations.

The authors attribute this to firstly the low hydraulic conductivity (unsaturated) of the B horizon, acting together with low root concentrations to produce lower soil conductances for water. For example, the time-averaged soil conductances, calculated for the B horizon and the topsoil for the period days 1-3 after irrigation, are 0.03 and 0.4 mm d^{-1} respectively. Secondly, less water is stored in the clay for each suction increment as a consequence of the steeper moisture characteristics of the clay (Olsson and Rose, 1978). The B horizon stores only 11.5 mm dm^{-1} depth or about 60% of that for an aggregated topsoil. Thus the topsoil and the B horizon to 400 mm depth dry to the wilting percentage at about the same time, but at markedly different rates of drying.

Many studies show higher root concentrations in clay subsoils after deep tillage (Mehanni, 1974; Taylor and Olsson, 1987), thereby potentially increasing the rate of water supply from the subsoil where previously limited by low root length (Hamblin, 1985). Passioura (1991), however, shows that while the root length per unit volume of soil might be higher on average, the roots may now be clumped in voids or between peds, with limited access to water stored in the clay matrix, resulting in little or no improvement in the rate of water supply to the crop. Further, even if the roots were to become evenly distributed throughout the dense fragments at high concentrations, both the low inherent water storage and the low hydraulic conductivity of the fragments (where these are unchanged from the parent clay) will restrict water supply. On such soils, yield increases after soil modification are attributable more to deeper wetting of the soil than to greater numbers of storage pores per se (Taylor and Olsson, 1987).

3. Root and Shoot Responses to Hard Subsoils

Plants grow more slowly in hard soil than in loose even when their roots apparently are able to supply adequate amounts of water and nutrients (Passioura, 1991). Hormonal mechanisms are suggested through which roots sense physical conditions in the soil, including possibly water supply, and communicate these to the shoot and modify the water relations and growth of the aerial parts (Masle and Passioura, 1987). Thus, while the overall extent of the root system determines the volume of water available to the crop, the size and vigor of the root system also controls the vigor of the aerial parts. Thus, any restriction on rooting imposed by either soil volume or soil conditions within the rooting volume, such as that imposed by a shallow, dense B horizon, is a limitation to crop size and vigor.

More commonly, roots cannot access water and nutrients at high rates in dense, non-porous clays. Many soils, even in beds, can have high macroporosities yet be hard in the matrix. Roots then grow slowly and slow top growth.

III. Improving and Managing Subsoil Structure

In this section we develop an approach to structural improvement of the subsoil that is based on meeting the requirements of the roots for their growth and functioning. The approach is illustrated by the B horizon of a Natrixeralf, Shepparton fine sandy loam (Skene and Poutsma, 1962). The soil typifies the problem over much of south-eastern Australia where slowly-permeable, sodic clays underlie shallow loam topsoils < 160 mm deep (McIntyre et al., 1982a,b).

We assume that fertilizers can supply the nutrients and that shading, together with mulching, can reduce extremes of soil temperature. We aim to restructure the clay B horizon, modifying its pore size distribution and mechanical strength to encourage continued root growth. We also aim to enhance supplies of water, oxygen and nutrients to the root surface by increasing the respective soil conductances and capacities, for a rapidly transpiring crop. We argue for a complete restructuring of the B horizon, aiming at a structure made up of stable, porous aggregates of low strength with macropores between. The first step is tillage to fragment the dense subsoil. However, the fragments produced are dense, angular, with low storage porosity, high strength and are unstable. So, the

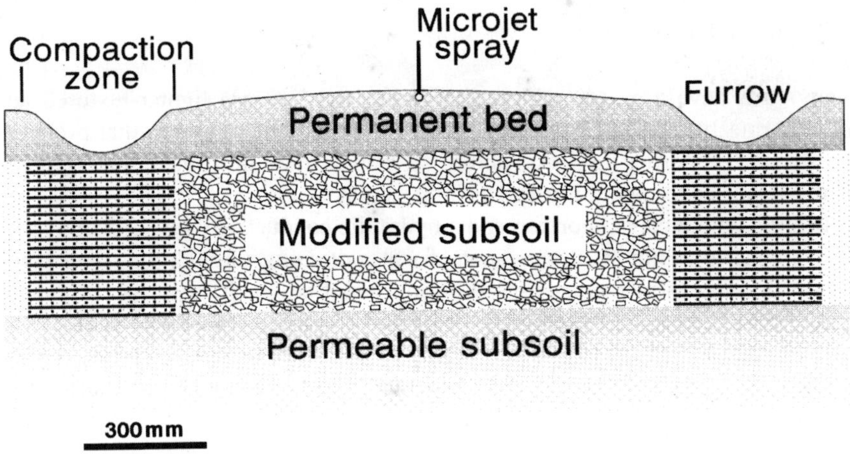

Figure 3. Illustration of a fragmented B horizon beneath a permanent (non-tilled) bed on the Shepparton soil showing 1) the permeable substratum, 2) dedicated lines for wheeled traffic and 3) controlled irrigation by microspray.

next step must be to convert these fragments into aggregates that are porous, rounded and stable, with high water storage and low strength.

Finally, we work towards new systems of soil preparation and management for farms, seeking to maintain long-term soil stability. Forming the topsoil up into beds (Cockroft and Tisdall, 1978; Tisdall and Hodgson, 1990) ideally allows one deep tillage operation to last indefinitely where subsequent cultivation of the bed and traffic compaction are eliminated and where the soil in the beds is wetted up slowly (Figure 3).

A. Field Specifications for Subsoil Structure

We assume the subsoil to be an environment for the growth and functioning of roots and that structure quality criteria suggested in the literature for topsoils can guide the restructuring of the subsoil to meet these requirements.

Each soil profile first needs to be assessed in relation to a set list of specifications before we can carry out the modification. We must ensure that any chemical constraints are eliminated and that the chemical requirements for the crop are met (Foy, 1992). While not within the scope of this paper, tillage of the subsoil gives the opportunity to incorporate any necessary ameliorants or fertilizer and, in addition, facilitates the leaching of toxins.

1. Depth of Tillage

Many soils have a profile with a clay B horizon between lighter-textured and more permeable A and C horizons (Figure 3). It is the B horizon that produces the adverse properties discussed above, so modification must be to the full depth of the B horizon, commonly to 600 mm depth.

In soils with a clay C horizon and a hydraulic conductivity lower than 60 mm d^{-1} we need to minimize the amount of water draining through the tilled layer. To do this the farmer must use additional practices such as surface drainage, spray rather than flood irrigation, pipe or mole drainage (MacEwan et al., 1992; Jayawardane and Chan, 1993), dewatering by winter cover crops, starting the wet season with a dry soil profile and deliberate mild throttling of infiltration with light compaction, crusting or slaking.

2. Fragment Size

The size distribution of fragments sought from deep tillage must facilitate the subsequent processes of conversion to porous aggregates. Our field experience indicates that we need to break up the subsoil into fragments, 80% being smaller than 20 mm, with an approximate size distribution: 20-10 mm, 40%, 10-2 mm, 40% and <2 mm, 20%. Biological agents find it difficult to mellow larger fragments quickly while fine material < 250 μm should not exceed 15% (Dexter, 1988). Misra et al. (1986a,b) show that unconfined aggregates smaller than about 20 mm can be readily penetrated by fine probes and roots. Specification of fragment size needs also to consider the dynamics of gas exchange in deep tilled soils (see Section III.D.4). Where gypsum is applied at tillage, fragment size is also determined by its rate of dissolution and penetration under the unsaturated conditions common to subsoils.

3. Gypsum Requirement

Sodic soils require gypsum to prevent dispersion by lowering exchangeable sodium and providing electrolyte (Shanmuganathan and Oades, 1983; Loveday, 1984). Gypsum addition prior to deep tillage helps the clay subsoil to loosen to a friable condition (Section III.B.2) and may allow effective tillage over a wider range of soil water content.

The amount of gypsum is specified by the Sodium Adsorption Ratio (SAR) and electrolyte values in the subsoil layer as described by Rengasamy et al. (1984). In Australian alfisols, gypsum is needed at a rate to lower the SAR (1:5 soil to water extract) to below 3 (Exchangeable Sodium Percentage of 6) and to provide an electrical conductivity of about 0.4 dS m^{-1} in the 1:5 soil to water suspension (Rengasamy and Olsson, 1991). Many soils do not need gypsum, because they are either naturally calcic or high in stabilizing agents such as sesquioxides.

4. Lime

Subsoils may need lime to correct acidity. Most nutritionists aim for a pH between 6.5 and 7.5. Lime also improves structural stability, providing bonds through cementation and through organic linkages (Oades, 1984). The deep tillage gives the opportunity to incorporate sufficient lime for a long-term benefit. In some cases, the intermixing of soil layers containing calcium minerals may be beneficial (Loveday, 1984; Jayawardane et al., 1994).

5. Stability

The soil must be stable to wetting, otherwise the fragments from the deep tillage would collapse again to a massive condition of low porosity and high strength. Our aim is to achieve a stability of Class 7 of Emerson (1967). Few Australian subsoils rate higher than Class 2, so considerable stabilization is usually needed. Once gypsum prevents dispersion, further stabilization involves the control of slaking and full stability may take several years (see Section III.C).

6. Macropores

Macropores include all those larger than 30 μm diameter. These pores conduct water quickly at suctions 0-10 kPa and are therefore important in infiltration, drainage (aeration) and the rapid redistribution of water. In addition, roots can grow in them (Russell, 1977; Jakobsen and Dexter, 1988). Most soil physicists specify that macropores should comprise at least 10% of the soil volume. In structurally-modified clays, however, with enhanced biological activity, 15-20% is more appropriate (Hall et al., 1977; Cockroft and Tisdall, 1978).

Oades (1984) uses the term "capillary pores" for those pores 25 to 100 μm diameter. These pores are important in the rapid redistribution of water under conditions of slow wetting of the beds by fine sprays and by capillarity from water ponded in the furrows.

The subsoil should also absorb water rapidly to reduce direct accessions to groundwater under conditions of ponded infiltration.

7. Storage Pores

Water is stored in pores between 30 and 0.2 μm diameter. As discussed in Section II.B.2, we need to increase the volume of available water stored in aggregates formed from the clay in order to be able to sustain high transpiration rates over time. This, ideally, would be a doubling of storage pores in the B horizon, thus avoiding the need for deeper modification to store and supply the same quantity of water. Soil scientists suggest that storage pores should comprise 20% of the soil volume (Hall et al., 1977).

8. Flow of Water to the Root

Water supply to roots requires both storage capacity and the ability of the soil to transmit water to the root surface. Water flows radially to the root surface through the storage pores. As for water storage, the main pores involved are 30 to 0.2 μm in diameter that comprise, ideally, at least 20% of the soil volume.

9. Soil Strength

Roots need a low soil strength for rapid growth and for the promotion of top growth (Passioura, 1991). Both the strength of the bulk soil and that of the clay aggregate need to be less than 1 MPa, as measured by a 2 mm conical-tip penetrometer. As shown in Figure 4, a penetrometer resistance of less than 0.5 MPa is necessary for unrestricted root growth (Cockroft et al., 1969; Taylor and Ratliff, 1969).

B. Fragmentation

Tillage can achieve an initial fragmentation of the dense layer (Spoor and Godwin, 1978; Jayawardane et al., 1988). We emphasize tillage as the necessary first step in the structure building process in the subsoil.

Of the various ways to fragment compacted subsoils, the rigid tine is both practical and effective. It is simple to operate, profile mixing can be controlled, the incorporation of ameliorants is easy and the equipment is cheap. In addition, the energy requirements are low compared with other methods.

Most efforts at deep tillage on farms are ineffective because the tines produce large clods at the soil surface and cut narrow slots in the subsoil with little shattering between (Trouse and Humbert, 1959; Blackwell et al., 1987). However, with the approach used by Spoor and co-workers (Spoor and Godwin, 1978, 1979; Spoor and Fry, 1983), tines break up the whole of the clay to the designated depth (Section III.A) and produce small fragments with little powder. Spoor and his co-workers in a series of excellent investigations have demonstrated the soil condition, implement design and method of operation needed.

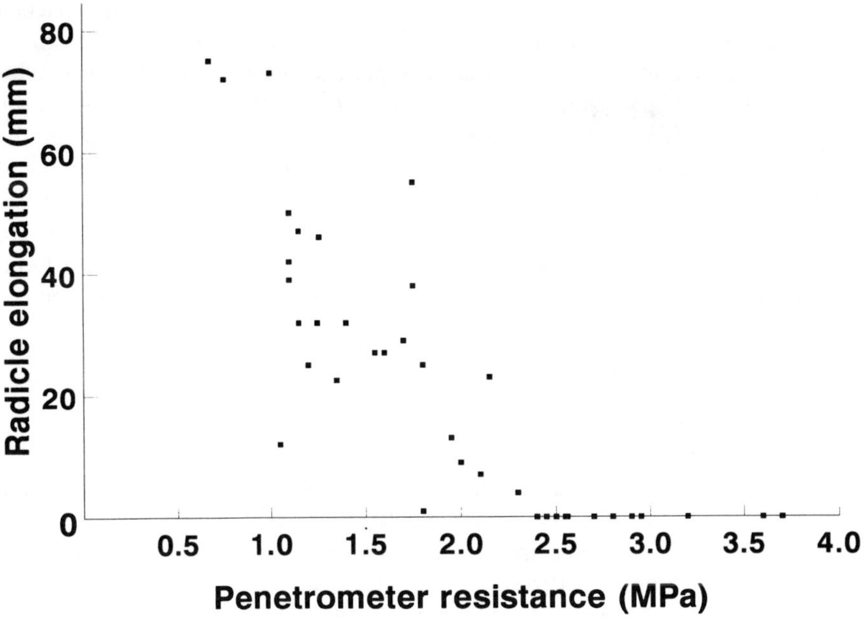

Figure 4. Elongation of pea radicles in relation to the penetrometer resistance of subsoil clay. (From Cockroft et al., 1969.)

1. Theory

Spoor and his co-workers have developed background soil mechanics theory for deep tillage with tines. As the tine moves forward, the soil fails by displacing a block forward, sideways and upwards along rupture planes from the tip of the tine (Payne, 1956). In addition, the tine forms fissures within the block and then the shank and point of the tine further fragment the soil as they force it around them. After every few centimeters of forward travel, the tine creates a new shear plane to the surface (Stafford, 1981). In this way the tine loosens the soil by shearing which is controlled by the confining pressures of both soil overburden and cohesion. It is the relationship between the shear stress by the tine and the confining pressure by the soil that influences the type of loosening that we achieve.

Spoor and Godwin (1979) used critical-state soil mechanics to relate soil volume (loosening or compaction) to shear stress and confining pressure. The tine must achieve brittle failure, and thus produce fragments, rather than plastic failure. We aim to have at least 80% of the fragments smaller than 20 mm (Section III.A.2). If the confining pressure is low during shear, the soil breaks up along shearing planes with brittle failure and the soil is loosened. However, if the confining pressure is high during shear, the soil compresses and fails on innumerable planes, rearranging the soil particles to an oriented, close packed structure. This is plastic failure.

Confining pressure is the important component to understand for successful deep tillage. Spoor and Godwin (1979) point out that many factors influence the magnitude of the confining pressures during deep tillage. The main ones are the magnitude of the resistance by the soil to shearing, the direction of loading and the strain imposed by the tine. Shearing resistance is affected by soil density, microstructure and water content; the direction of loading and the strain are affected by the geometry of the implement and its operation. So, in the field, effective deep tillage with tines depends on the condition of the soil, implement design and implement operation.

2. Soil Condition

Soil water content is important (Spoor and Godwin, 1979) because the tine shatters dry soil into clods and powder and deforms wet soil into smeared, plastic lumps. In both cases, tractor draft is high. The ideal water content is just drier than the Plastic Limit (Stafford, 1979), approximated in the field when the operator can roll the soil into a 3 mm diameter rod, but only with some difficulty.

The type of clay mineral influences plasticity. Illitic and kaolinitic clays fragment more readily than smectitic whose clods may need mellowing by wetting and drying cycles. Mechanical composition is important; loams are more easily fragmented than clays, for instance. The nature of the physico-chemical properties is crucial. Dispersive clays are much less friable (Shanmuganathan and Oades, 1982). These need gypsum well before deep tillage so that rain and irrigation can transport the gypsum into the clay. On the Shepparton soil, the upper B horizon, while non-sodic, has a requirement for electrolyte to prevent dispersion under mechanical action (Class 2a of Rengasamy et al., 1984). This means that reacting gypsum with the entire matrix to the designated depth before tillage may be beneficial; flocculation of the clay then facilitates the desired fragment-size distribution and requires low energy input during tillage.

Any other physico-chemical property that leads to brittle failure can be important; these include organic matter, sesquioxides, electrolytes and cementation bonding.

The operator can judge the condition of the soil by taking a 20 mm piece of soil at the Plastic Limit and applying a compressive shear force exerted between the thumb and fingers (Butler, 1955). Subsoils that shear into peds are suitable for deep tillage. Subsoils that coalesce into plastic rods and balls need treatment if possible; thus, a plastic, sodic subsoil can become friable when pre-treated with gypsum.

3. Implement Design

In our aim of brittle failure to small fragments, the implement must fail the soil with an upward tendency rather than forward. By failing towards the ground surface, the smaller pressures achieve more brittle failure and the tractor uses less power. This dictates a low angle of tine point to the horizontal and Spoor and Godwin (1978) suggest 20°.

Increasing the tine width gives more brittle failure because wider tines increase the strain and therefore the chance to shear.

Wings attached to the bottom of the ripper increase the width ripped, increase the depth the operator can achieve at each pass and increase the amount of fragmentation. Once the point of the tine breaks up the soil by crescent failure and fissuring, the fragments and clods are sheared by the inclined (20°) face of the wing under the light confining pressure of the overburden; remaining pieces pass over the top of the wing and fail in tension (Spoor and Godwin, 1978).

Tine spacing influences the shattering because when wide apart they cannot break out the soil between. Tines should be spaced at 0.5 to 0.75 times the working depth.

4. Operation

For any tine there is a critical working depth and if the operator penetrates the tine deeper he will cause compaction rather than loosening (Spoor, 1979). When operating deeper than the critical depth, the tine cannot fail the soil to the ground surface and it compacts the surrounding soil to accommodate the tine's passage. The operator will have to make more than one pass to remain above the critical depth. The first pass fails the soil to the ground surface and in deeper passes the tine fails the soil only to the depth of the previous loosening. It is better still if he attaches shallower tines to the toolbar ahead of the deeper tines. Critical depths are commonly 300 to 400 mm. The operator can check whether the tine is below critical depth with a small trench across the direction of travel, looking for shattering to the working depth between tines, in contrast to narrow slots only. Spoor and Godwin (1978) achieved considerable increases in soil disturbance with no increase in power required when they used leading tines.

The speed of the tractor affects the confining stresses. Stafford (1981) showed that the type of failure around the tine can change from brittle to plastic as speed increases. This is because the higher speed builds up higher confining stresses. Increasing speed also increases the force required to accelerate the failed soil and so increases the draft.

C. Stabilization

In this section, we apply well established principles of soil stabilization, developed mainly for topsoils, to the problem of managing tilled subsoils beneath beds. Little is evident in the literature, however, where this has been attempted systematically. In this section we discuss the physical, chemical and biological agents that produce stability.

The loose assortment of calcic subsoil fragments the ripper produces has no inherent stability or organized structure. The immediate problem the farmer faces is that the clayey subsoil is highly unstable to wetting; that is, the soil may slake or slump at low matric suction. We assume that the surface soil is stable or can be made stable by minimum cultivation, high organic matter and root and other biological activity. We also assume that the farmer will try to avoid compaction of the soil and will aim for minimum cultivation by using beds.

1. Chemical Ameliorants

We have already discussed the initial application of ameliorants such as gypsum and lime (Section III.A). However, the farmer can apply additional gypsum as necessary over time to maintain electrolyte and low exchangeable sodium.

2. Roots

The root is probably the most important agent in stabilizing the tilled subsoil. The low strength of the tilled subsoil provides the opportunity for roots to grow to high concentrations. However, we must take steps immediately to prevent soil structural collapse, particularly on wetting. Our experience shows that roots will reduce this if grown immediately. The roots ramify between the fragments and particles and keep them separate from one another and so act to minimize collapse. This structure is fragile but should be sufficient to keep the soil open until the soil develops long-term stability.

Lasting stability comes from the long-term process that involves build up of aggregates from fine material (Tisdall and Oades, 1982). Roots exude mucigels that bond particles and supply organic matter that will bond particles within microaggregates (< 250 μm); they also host fungi and other microorganisms. These processes occur mostly within the rhizosphere.

Roots and fungal hyphae will also exploit any opportunity to penetrate the clay fragments. They can do this because some pores are large enough for fine roots to penetrate (Hamblin, 1985), and because the clay fragments are smaller than 20 mm and are unconfined, they will more readily fail (Misra et al., 1986a,b). Root entry into fragments is shown by results from a greenhouse experiment at Tatura, where fragments, 4-6 mm, of sodic clay were wetted slowly under the unsaturated conditions produced by a throttle layer of fine soil at the soil surface.

Table 2. Penetration of unconfined fragments of sodic and calcic clay by ryegrass roots[a]

	Ryegrass roots	
	Mass (mg)	Length (mm)
Ryegrass - gypsum	8.9	128
Ryegrass + gypsum	18.5	223
LSD (P=0.05)	5.9	81

[a] In 5 g air-dry fragments, 4-6 mm diameter.
(From Olsson and Rengasamy, unpublished data.)

Over a five-week growth period the roots of ryegrass entered the unconfined fragments (Table 2). Unsown treatments, with and without gypsum addition, were included in the experiment to enable correction for background roots. Where ryegrass was grown, the fragments treated with gypsum, mixed dry at packing, supported almost twice the length and mass of ryegrass roots. The increased root entry into the gypsum-treated fragments suggests a lower resistance to root penetration.

The stability in water of the fragments from this experiment was reported by Rengasamy and Olsson (1991). Clay dispersion was completely controlled in the gypsum-treated fragments under the experimental conditions of unsaturated wetting of the clay (Table 3). However, ryegrass roots were ineffective in controlling dispersion over the study period without gypsum treatment to remove sodium. Stability in water was measured without mechanical agitation. The percentage of water stable particles > 2 mm, while high in all treatments, was significantly increased by ryegrass, only where gypsum was added.

To ensure that the soil stays loose, it is essential to encourage high concentrations of roots and to maintain these year round. Oades (1984) notes that fallow, a period with no plants growing, is detrimental to the structure of surface soils. We thus aim to keep the roots growing continuously using successive crops and by keeping the soil moist and fertilized.

3. Fungi

Fungi could be almost as important as roots in soil stabilization because although they are smaller they are far more numerous (Dexter, 1991). These agents are either symbiotic with the roots of higher plants or saprophytic on organic matter in the soil. They are active in binding microaggregates (< 250 μm), extending the effectiveness of roots into this range (Tisdall and Oades, 1982; Foster et al., 1983). Fungal hyphae can penetrate fragments by exerting axial pressures of the same order as roots (R.C. Foster, personal communication). We aim for high

Table 3. Effect of gypsum and ryegrass on clay dispersion and stability of fragments in water

Treatment	Spontaneously dispersed clay %	Water stable fragments >2 mm % (w/w)
Unsown control	10.5	71.4
Unsown + gypsum	0.8	77.1
Ryegrass - gypsum	12.1	70.1
Ryegrass + gypsum	0.9	82.1
LSD (P=0.05)	2.7	4.4

(From Rengasamy and Olsson, 1991.)

concentrations of hyphae by providing sufficient energy, keeping the soil moist and managing nutrients.

4. Other Microorganisms

Bacteria, actinomycetes and other microorganisms are also important in stabilizing soil through their exudates and products of decomposition. We can maintain high populations by ensuring substrate and water.

5. Water

Water affects stability in two ways; through the rate of wetting and through effective stress. Rapid wetting produces slaking by differential swelling, air entrapment and heat of wetting. The adverse effects of slaking can be overcome by slow wetting, by avoiding complete drying before irrigation or by adding organic matter. Soil-water suction has an effect when the soil is being wetted and during the drying phase; whenever soil-water suction is close to zero, effective stress is low and the soil unstable (Mullins et al., 1990). Management of water is thus fundamental to the stabilization of soils.

6. Drying

When we dry a soil and as the water menisci retreat, the soil particles come closer together; eventually particle edges come close enough for various bonds to form and for cohesion to develop (Dexter, 1988). When we need to aggregate fine soil, wetting must be followed by drying to develop the aggregation.

7. Organic Matter

Organic matter is essential to stabilize most soils (Russell, 1973) but most subsoils need large quantities. Organic matter is the source of the bonds that stabilize particles (Emerson, 1967). It also is the source of energy for the soil organisms that also stabilize soils. It also forms films and protects the soil particles from rapid wetting (Greenland, 1981; Rengasamy and Olsson, 1991). Organic matter also acts as a general cement in the form of mucigel. In the subsoil, we manage organic matter by no subsequent cultivation to the original deep tillage and by ensuring that plenty of roots grow.

8. Earthworms

Earthworms stabilize soils by intimate mixing of stabilizing agents, such as organic matter, in the gut. They also transport stabilizing agents and stabilized surface soil into the subsoil (Russell, 1973).

D. Porosity

As described in Section II, one of the major problems of the natural subsoil is its low porosity in the macropore and storage pore classes. The initial fragmentation and the subsequent management aim to enhance all processes of aeration and water supply (see Section III.A). The first step is to increase the volume of macropores by the fragmentation. Then we aim to increase the volume of storage pores within the fragments (Section III.A.7) and at the same time ensure their stability (Section III.C). These stable, porous fragments we now call aggregates.

1. Infiltration

The pores involved in ponded infiltration include cracks, earthworm tunnels, larger root channels and pores formed by tillage (Klute, 1982; Hamblin, 1985; Jayawardane and Prapather 1992). The water flows in pores larger than 30 micron diameter. Cracks form naturally in clay soils, earthworms produce pores when the soil is mulched, but roots are effective only after they decay; tillage is effective only when the soil is stable in water.

In soils that are stable in water, high infiltration rates are possible. However, many surface soils are not stable and are susceptible to rapid wetting and water saturation, both of which cause compaction. In unstable soils, slaking from rapid wetting can be prevented by low-rate sprinklers or a throttle at the infiltrating surface.

Increased macroporosities have been reported in clay subsoils in deep tilled and slotted systems (Mehanni, 1974; Blackwell et al., 1990; Jayawardane and Chan, 1993), leading to improved infiltration and drainage in the short term. The increased porosity can be soon lost under flood irrigation and where traffic compaction is allowed (van Ouwerkerk and Raats, 1986). There are few data, however, on the persistence of these pores in modified subsoils over periods longer than 1-2 years. Under spray irrigation, Taylor and Olsson (1987) showed that lucerne yields were increased with added gypsum while gypsum plus ripping to 600 mm depth resulted in higher macroporosity and yield over a four-year experimental period. The yield increase was closely related to water infiltrated at each irrigation and hence the macroporosity of the B horizon. Under flood irrigation, however, gypsum plus ripping did not increase yield above that for gypsum singly. This was attributed to the detrimental effect of flooding on the stability of soil pores since traffic compaction was confined to times when the soil was dry, during haymaking. A similar conclusion was reached by Loveday et al. (1970).

2. Redistribution of Water

In beds, capillary phenomena predominate during infiltration and the redistribution of water (Section III.A.6). The slow application rate ensures that the B horizon is always wetted under suction. With time, these pores remain functional because of the control of soil-water suction and the activity of the biological agents.

3. Drainage

We assume an unrestricted path for water flow through the modified B horizon under gravity. Water flows through the pores of greater than 30 μm diameter that are involved in infiltration and redistribution.

4. Aeration

Our aim, for the Shepparton soil is an aggregated soil to some 600 mm depth (Figure 3). Gas transport by diffusion and mass flow occurs in the air-filled pores. On drainage, all fragments to 600 mm depth in the modified soil are initially surrounded by air space. Smaller fragments are aerated within (Currie, 1962; Greenwood, 1975). Larger fragments, although they exchange gases slowly because of their low porosity, do not need high rates initially because of their low biological activity and hence, low oxygen consumption. As the fragments develop more biological activity over time, their porosity increases and gas transport into the aggregate increases.

Tisdall et al. (1984) describe fluctuations in oxygen concentration over depth and time beneath high-yielding peach trees growing on 9-year old beds in which the topsoil is hilled to 370 mm deep. Over a 5-day period following irrigation by sprays, the lowest oxygen concentration recorded in the upper 800 mm of soil, 11% by volume, remained higher than the 6% concentration suggested by Greenwood (1975) as the lower limit for root growth. In the B horizon, which had been ripped and treated with gypsum nine years previously, values of air-filled porosity were 7-9% by volume over the period.

5. Water Storage

The pores involved in the storage of water are of 30 to 0.2 μm diameter (Table 1). The agents that produce these pores include fungal hyphae, lateral roots, root hairs, soil swelling and shrinking (Hamblin, 1985; Dexter, 1988, 1991). In addition, the processes of building aggregates out of finer material produces pores of this size. In fragments larger than 250 μm, within the macroaggregate size range, we aim to double the complement of storage pores in the B horizon. We manage the fungi by providing energy, water and aeration, the energy coming from organic matter in the soil (R.C.Foster, personal communication) and in the case of VAM fungi, from the host root (Tisdall and Oades, 1979, 1982). Fungal hyphae can then grow into the fragments producing pores 0.5 to 2 μm in diameter. Fine roots and root hairs (50 to 5 μm diameter) will penetrate the dense fragments when grown at high concentrations. We can assume that swelling and shrinkage creates pores because the soil is calcic.

In fragments smaller than 250 μm, the aim is to combine these into macro-aggregates as described by Tisdall and Oades (1979,1982).

When repacked into rhizotron compartments (2.5 m^3) at the Tatura Institute, sodic clay fragments (< 25 mm) from the B horizon of the Shepparton soil, when mixed dry with natural gypsum (0.47% w/w soil), showed the development of new storage pores within the fragments over a 40-week period during which crops of irrigated soybean and wheat were grown (K.A. Olsson, P. Willoughby and J.J. Millar, unpublished data). Scanning electron micrographs (Plate 1) of medial fracture surfaces of representative fragments show the development of porosity, including storage pores, in fragments treated with gypsum, in contrast to untreated fragments.

The increase in porosity was detectable at the level of the bulk soil as an increase in hydraulic conductivity, measured in situ, at matric suctions over the range 1.0 to 39 kPa. Regression analysis of the log hydraulic conductivity-log matric suction characteristics of the soils over this range of suction showed the gypsum-treated subsoil to have a significantly lower slope ($b = -2.39$, $r^2 = 0.73$) to that of the untreated subsoil ($b = -4.65$, $r^2 = 0.75$). Hydraulic conductivities at 39 kPa suction are 2.0×10^{-4} and 6.1×10^{-3} mm d^{-1} for the non-gypsum and gypsum-treated clays respectively. These data support the SEM observations in showing an increase in porosity in the range of pore diameters down to at least 7 μm during cropping and following gypsum treatment. Such changes in microstructure are consistent with gypsum flocculating the clay and from the

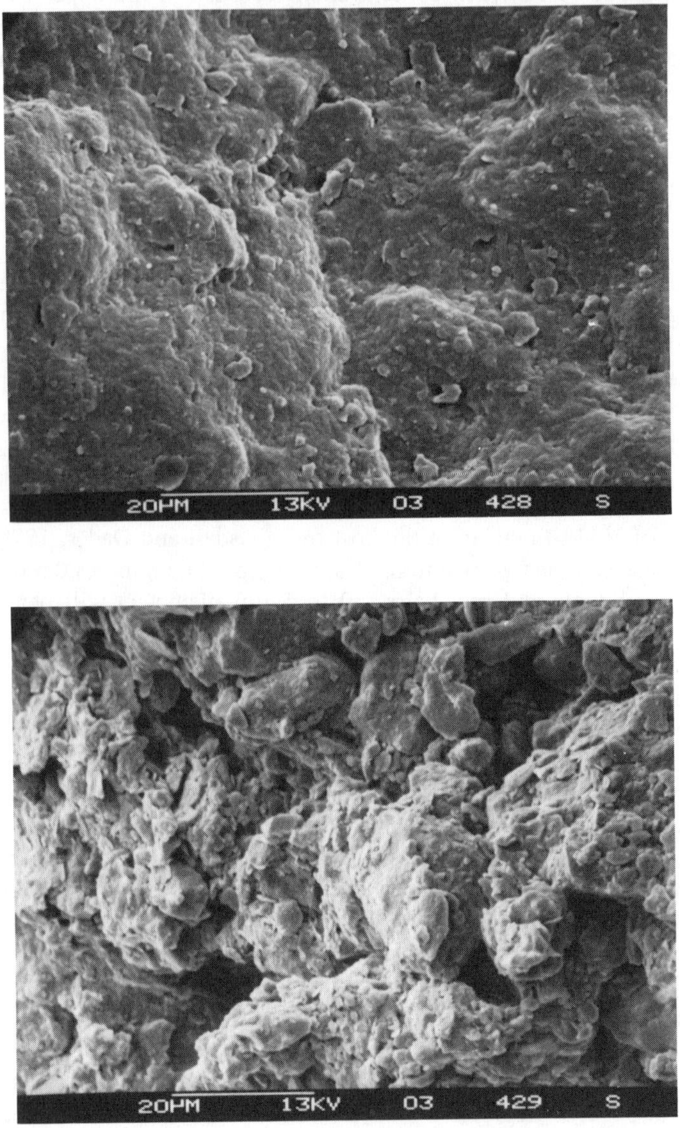

Plate 1. Scanning electron micrographs of medial fracture surfaces of clay fragments sampled from rhizotron compartments after 40 weeks of irrigated cropping. Regions within 1-2 mm of the edges of the fragments are compared for: top) Sodic clay (no gypsum applied); bottom) Gypsum-treated.
(From Olsson, Willoughby, and Millar; unpublished.)

enhanced penetration of fine roots and other biological activity in the rhizosphere under continuous cropping.

6. Water Flow to the Root

Water flows to the root in pores of 30 to 0.2 μm diameter that have a direct contact connection to the surface of the root. The agents responsible for their production and their management are the same as those described for storage pores.

E. Strength Reduction

Once the subsoil is fragmented, crop roots can grow in the macropores between the fragments. However, the roots may become confined and reduce top growth (Passioura, 1991). In addition, we aim for higher root concentrations so that we need roots to penetrate the fragments to stabilize them. So, we have to reduce the strength of the fragments. This complements the need to increase the porosity for water storage. A reduction in strength of fragments is feasible because they are in a calcic, unconfined, moist and well-drained environment.

We suggest that there are three ways which we could use to reduce the strength of subsoil fragments:

1. Dilation

Dilation by shearing is a consequence of the action of the tine at the original deep tillage. The tine first produces fragments and then shears many of them under confining pressure from the overburden. The shearing breaks the bonds that hold the clay particles together in their cohesive state; these then dilate to a less dense state. Clearly, the dilated fragment must be stabilized as soon as possible.

2. Wetting and Drying

Wetting and drying cycles can reduce the strength of fragments because the lack of confinement produced by the deep tillage permits the clay fragments to mellow (Dexter, 1988). When clays are wetted and are free to swell, subsequent drying forms planes of weakness. Successive cycles create planes between successively smaller units thoughout the fragment. Organic matter including macro-organic matter is necessary to prevent the units from coalescing again.

3. Biological Activity

Biological activity is the most likely way in which we can reduce the strength of fragments and, in addition, it is essential in stabilizing them and in converting fragments into aggregates. Biological agents include the roots, soil animals, fungi and other microorganisms we describe in Section III.D. They can exploit the reduction in strength caused by dilation and cycles of wetting and drying. Cockroft et al. (1969) show the reduction in the strength of a clay subsoil at a given water content from increasing the porosity.

The roots and fungal hyphae produce pores that weaken the large fragment and progressively produces more pores until a strength less than the desired 0.5 MPa penetrometer resistance is eventually reached. This may take several years. We describe the management needed in Section III.D.

When a root attempts to penetrate a dense fragment, the pressure it must develop to penetrate builds up and may stop the root growing within a few millimeters. The root then grows parallel to the surface of the fragment, failing the overlying soil to the unconfined surface. This weakens the dense mass so that successive roots penetrate the fragment more deeply, failing the soil into the weakened layer above it.

IV. Soil Management Systems for Crops

In this section we set out the basic requirements for soil management systems as they apply to different crops. We describe soil management systems for fruit trees, row crops, and field crops. The development of structure in the subsoil is best achieved in beds with controlled irrigation and where traffic compaction is eliminated by confining wheeled traffic to dedicated traffic lines. Each system, compared with conventional, must be high yielding, simple, and sustainable.

A. Fruit Trees

The grower must first have his soil examined to assess soil properties such as depth to till, gypsum requirement and pH. He then grades the land surface to the slope needed for good surface drainage. He then provides the nutritional needs of the trees, incorporating those immobile nutrients as necessary by the subsequent deep tillage. He must apply whatever gypsum is needed well before deep tillage. This ensures that the failure of the clay is brittle and not plastic (see Section III.B). He waits until the soil reaches the Plastic Limit, then deep tills, but only below the proposed tree lines and not the traffic lanes between. He should dig several trenches to assess the fragmentation. He then cultivates the traffic lane to the full depth of the topsoil and forms this loose soil up into beds. All future traffic is confined to the traffic lanes.

He now practices zero tillage. He immediately sows a stabilizing crop, usually ryegrass in temperate regions, encouraging rapid growth to minimize collapse of the loose soil. He slashes the stabilizing crop as necessary. To prevent soil collapse he must grow the stabilizing crop year round until the tree roots occupy the whole of the bed. Competition by the stabilizing crop can be avoided by herbicide spraying around the trees. When the bed is fully permeated by tree roots, within perhaps two years, the stabilizing crop is necessary only during the dormant period when the tree roots are not growing.

Growing the stabilizing crop is essential to control slumping, but so is the management of water. This involves slow wetting and good drainage. The orchardist achieves slow wetting by using spray irrigation at application rates lower than four mm per hour. Low intensity rain also meets these criteria; the soil is protected from heavy rain by the tree canopy, the mulch from the stabilizing crop and by the slope of the bed surface. Good drainage is ensured by the deep tillage. Further, the orchardist must match his irrigation water to the requirements of the trees to avoid any excess.

The orchardist irrigates and fertilizes his trees for maximum yields – keeping the soil wetter than about 100 kPa suction and leaf nutrients at recommended levels. This system of soil management enables the orchardist to manage all five root requirements (Section II.B) and thus promote root growth.

B. Row Crops

The row crop farmer can use the same general approach to soil preparation and management. He examines his soil, grades it, then adds the necessary fertilizer and gypsum. Next, he deep tills, cultivates and forms up beds. Clearly, the bed width is made to fit tractor and implement wheel spacings and all traffic travels in the furrows. He sows his crop immediately, thus avoiding any fallow period, so that the roots keep the soil loose. As soon as he harvests the crop he must aim to grow another crop to maintain soil stability through roots. He now practices zero tillage. If the farmer irrigates by sprinkler, he takes the same precautions as the orchardist. If he does not spray, he must irrigate by furrow and not flood his soil. He forms his soil up into beds and runs water along the furrows. He reduces the rate of wetting in the bed and avoids saturation by using a throttle at the furrow face by slaking the surface layer of soil. He irrigates and fertilizes appropriately.

C. Field Crops

The farmer can use the same approach. However, high beds and deep furrows may be difficult for him to manage and the relief of the bed may have to be low. The important point is that all traffic is confined to the furrow base to avoid any re-compaction of the soil structure. The furrows provide for drainage of surface water and, if necessary, irrigation.

V. Summary

The current state of structure modification on dense clay soils has not achieved the productive potential of the world's best soils. Subsoils with poor structure, especially where compacted, constrain crop productivity because they restrict the development and functioning of root systems. This paper describes how to restructure these subsoils by deep tillage and how to manage them subsequently to maintain high yields from crops.

First, we describe the limitations to yield imposed by poorly structured subsoils, including water infiltration, drainage, water flow to the root and mechanical resistance. We then set up specifications for the restructured soil such as depth required and its stability, porosity and strength. We discuss the factors important in successful deep tillage including background theory, the initial soil condition, implement design and operation. Stabilizing the loose subsoil involves several agents including chemical ameliorants, grass roots, fungi, other microorganisms, water management, organic matter and earthworms.

The paper then describes how to develop and manage soil porosity, including pores that influence infiltration, drainage, aeration, water storage and water flow to the root. Next it suggests how the high strength of the fragments produced by the deep tillage can be reduced by dilation during the initial tillage, subsequent wetting and drying cycles, chemical action, and biological action over time. Finally, we suggest systems of soil management for fruit trees, row crops and field crops, by which the farmer could use permanent, non-tilled beds to produce sustained high yields.

Acknowledgments

The authors thank K. Samarasinghe for technical assistance, K.P. Robinson for preparing the figures and the Murray-Darling Basin Commission, Natural Resources Management Strategy, for funding part of the work referred to in this paper.

References

Bergman, H.F. 1959. Oxygen deficiency as a cause of disease in plants. *Bot. Rev.* 25:418-485.

Blackwell, P.S., T.W. Green, and K.A. Olsson. 1987. The size and horizon-of-origin of fragments produced by deep ripping texture contrast soils. *Aust. J. Soil Res.* 25:211-222.

Blackwell, P.S., A.J. Ringrose-Voase, N.S. Jayawardane, K.A. Olsson, D.C. McKenzie, and W.K. Mason. 1990. The use of air-filled porosity and intrinsic permeability to air to characterize structure of macropore space and saturated hydraulic conductivity of clay soils. *J.Soil Sci.* 41:215-228.

Boone, F.R. 1988. Weather and other environmental factors influencing crop responses to tillage and traffic. *Soil Tillage Res.* 11:283-324.

Butler, B.E. 1955. A system for the description of soil structure and consistence in the field. *J. Aust. Inst. Agric. Sci.* 21:239-249.

Cannell, R.Q., and M.B. Jackson. 1981. Alleviating aeration stresses. p. 141-192. In: G.F. Arkin and H.M. Taylor (eds.) *Modifying the root environment to reduce crop stress.* ASAE, Michigan, WI.

Cockroft, B., and F.M. Martin. 1981. Irrigation. p.133-147. In: J.M. Oades, D.G. Lewis, and K. Norrish (eds.) Red-brown earths of Australia. University of Adelaide and CSIRO, Division of Soils, Adelaide, South Australia.

Cockroft, B., and J.M. Tisdall. 1978. Soil management, soil structure and root activity. p. 387-391. In: W.W. Emerson, R.D. Bond and A.R. Dexter (eds.) *Modification of soil structure.* Wiley, Chichester.

Cockroft, B., and J.C. Wallbrink. 1966. Root distribution of orchard trees. *Aust. J. Agric. Res.* 17:49-54.

Cockroft, B., K.P. Barley and E.L. Greacen. 1969. The penetration of clays by fine probes and root tips. *Aust. J. Soil Res.* 7:333-348.

Currie, J.A. 1962. The importance of aeration in providing the right conditions for plant growth. *Journal of the Science of Food and Agriculture.* 13:380-385.

Dexter, A.R. 1988. Advances in characterization of soil structure. *Soil Tillage Res.* 11:199-238.

Dexter, A.R. 1991. Amelioration of soil by natural processes. *Soil Tillage Res.* 20:87-100.

Eck, H.V. and P.W. Unger. 1985. Soil profile modification for increasing crop production. In: B.A. Stewart (ed.) *Adv. Soil Sci.* 1:65-100.

Emerson, W.W. 1967. A classification of soil aggregates based on their coherence in water. *Aust. J. Soil Res.* 5:47-57.

Flett, S.P. 1986. *Phytophthora nicotianae var. nicotianae* causing root and crown rot of direct-seeded tomatoes in Victoria. *Australasian Plant Path.* 15:11-13.

Foster, R.C., A.D. Rovira, and T.W. Cock. 1983. *Ultrastructure of the root soil interface.* Am. Phytopathol. Soc., St. Paul, Minnesota.

Foy, C.D. 1992. Soil chemical factors limiting plant root growth. In: J.L. Hatfield and B.A. Stewart (eds.) *Adv. Soil Sci.* 19:97-149.

French, R.J. and J.E. Schultz. 1984. Water use efficiency of wheat in a Mediterranean-type environment. II. Some limitations to efficiency. *Aust. J. Agric. Res.* 35:765-775.

Gardner, W.R. 1960. Dynamic aspects of water availability to plants. *Soil Sci.* 89:63-73.

Greacen, E.L. 1977. Mechanisms and models of water transfer. p. 163-196. In: J.S. Russell and E.L. Greacen (eds.) *Soil factors in crop production in a semi-arid environment.* Univ. of Queensland Press and Aust. Soc. Soil Science Inc., St. Lucia, Q'ld, Australia.

Greenland, D.J. 1981. Soil management and soil degradation. *J. Soil Sci.* 32:301-322.

Greenwood, D.J. 1975. Measurement of soil aeration. p. 261-272. In: *Soil physical conditions and crop production.* Tech. Bull. 29. Min. Agric. Fish. Food, H.M.S.O., London.

Gregory, P.J., J.V. Lake and D.A. Rose. (eds.) 1987. *Root development and function.* Cambridge University Press, London.

Hall, D.G.M., M.J. Reeve, A.J. Thomasson and V.F. Wright. 1977. Water retention, porosity and density of field soils. Soil Survey Technical Monograph No.9, Rothamsted, Harpenden.

Hamblin, A.P. 1985. The influence of soil structure on water movement, crop root growth, and water uptake. In: N.C. Brady (ed.) *Adv. Agron.* 38:95-158.

Jakobsen, B.F., and A.R. Dexter. 1988. Influence of biopores on root growth, water uptake and grain yield of wheat *(Triticum aestivum)* based on predictions from a computer model. *Biol. Fertil. Soils.* 6:315-321.

Jayawardane, N.S. and W.S. Meyer. 1985. Measuring air-filled porosity changes in an irrigated swelling clay soil. *Aust. J. Soil Res.* 23:15-22.

Jayawardane, N.S. and S.A. Prathapar. 1992. Effect of soil loosening on the hydraulic properties of a duplex soil. *Aust. J. Soil Res.* 30:959-975.

Jayawardane, N.S. and K.Y. Chan. 1994. The management of soil physical properties limiting crop production in Australian sodic soils – a review. *Aust. J. Soil Res.* 32:13-44..

Jayawardane, N.S., J. Blackwell, and P.S. Blackwell. 1988. Fragment size distribution within slots created in a duplex soil by a prototype rotary slotter. *Soil Tillage Res.* 12:53-64.

Jayawardane, N.S., J. Blackwell, G. Kirchhof, and W.A. Muirhead. 1994. Slotting – a deep tillage technique for ameliorating sodic, acid and other degraded subsoils and for land treatment of waste. *Adv. Soil Sci.* (in press).

Klute, A. 1982. Tillage effects on hydraulic properties of a soil: a review. p. 29-43. In: P.W. Unger, D.M. van Doren, F.D. Whisler, and E.L. Skidmore (eds.) *Predicting tillage effects on soil physical properties and processes.* Amer. Soc. Agron. Special Publication No. 44. Amer. Soc. Agron. and Soil Sci. Soc. Amer., Madison, Wisconsin.

Letey, J. 1985. Relationship between soil physical properties and crop production. In: B.A. Stewart (ed.) *Adv. Soil Sci.* 1:277-291.

Loveday, J. 1984. Amendments for reclaiming sodic soils. p. 220-227. In: I. Shainberg and J. Shalhevet (eds.) *Soil Salinity under Irrigation: Processes and Management.* Springer-Verlag, Berlin.

Loveday, J., J.E. Saunt, P.M. Fleming, and W.A. Muirhead. 1970. Soil and cotton responses to tillage and amelioration treatments in a brown clay soil. I. Soil responses and water use. *Aust. J. Exp. Agric. Anim. Husb.* 10:313-324.

MacEwan, R.J., W.K. Gardner, A. Ellington, D.G. Hopkins, and A.C. Bakker. 1992. Tile and mole drainage for control of waterlogging in duplex soils of southeastern Australia. *Aust. J. Exp. Agric.* 32:865-878.

McIntyre, D.S., J. Loveday, and C.L. Watson. 1982a. Field studies of water and salt movement in an irrigated swelling clay soil. I. Infiltration during ponding. *Aust. J. Soil Res.* 20:81-90.

McIntyre, D.S., J. Loveday, and C.L. Watson. 1982b. Field studies of water and salt movement in an irrigated swelling clay soil. II. Profile hydrology during ponding. *Aust. J. Soil Res.* 20:91-99.

Marshall, T.J. 1959. Relations between water and soil. Tech. Comm. No.50. C'wealth Agric. Bureaux, Harpenden, U.K.

Masle, J. and J.B. Passioura. 1987. The effect of soil strength on the growth of young wheat plants. *Aust. J. Plant Physiol.* 14:643-656.

Mehanni, A.H. 1974. Short term effect of some methods for improving soil structure in red-brown earth soils of the northern irrigation areas, Victoria. *Aust. J. Exp. Agric. An. Husb.* 14:689-693.

Meyer, W.S. and H.D Barrs. 1991. Roots in irrigated clay soils: measurement techniques and responses to rootzone conditions. *Irrig. Sci.* 12:125-134.

Meyer, W.S., H.D. Barrs, R.C.G. Smith, N.S. White, A.D. Heritage, and D.L. Short. 1985. The effect of irrigation on soil oxygen status and root and shoot growth of wheat in a clay soil. *Aust. J. Agric. Res.* 36:171-185.

Misra, R.K., A.R. Dexter, and A.M. Alston. 1986a. Penetration of soil aggregates of finite size. I. Blunt penetrometer probes. *Plant and Soil.* 94:43-58.

Misra, R.K., A.R. Dexter, and A.M. Alston. 1986b. Penetration of soil aggregates of finite size. II. Plant roots. *Plant and Soil.* 94:59-85.

Mullins, C.E., D.A. MacLeod, K.H. Northcote, J.M. Tisdall, and I.M. Young. 1990. Hardsetting soils: behaviour, occurrence and management. In: R. Lal and B.A. Stewart (eds.) *Adv. Soil Sci.* 11:37-108.

Oades, J.M. 1984. Soil organic matter and structural stability: mechanisms and implications for management. *Plant and Soil.* 76:319-337.

Olsson, K.A. and C.W. Rose. 1978. Hydraulic properties of a red-brown earth determined from in situ measurements. *Aust. J. Soil Res.* 16:169-180.

Olsson, K.A. and C.W. Rose. 1988. Patterns of water withdrawal beneath an irrigated peach orchard on a red-brown earth. *Irrig. Sci.* 9:89-104.

Passioura, J.B. 1991. Soil structure and plant growth. *Aust. J. Soil Res.* 29:717-728.

Payne, P.C.J. 1956. The relationship between the mechanical properties of soil and the performance of simple cultivation implements. *J. Agric. Engng. Res.* 1:23-50.

Penman, H.L. 1971. Water as a factor in productivity. p. 89-99. In: P.F.Wareing and J.P. Cooper (eds.) *Potential crop productivity.* Heinemann, London.

Rab, Md.A. and S.T. Willatt. 1987. Water use by irrigated potatoes on a duplex soil. *Aust. J. Exp. Agric.* 27:165-172.

Rengasamy, P. and K.A. Olsson. 1991. Sodicity and soil structure. *Aust. J. Soil Res.* 29:935-952.

Rengasamy, P., R.S.B. Greene, G.W. Ford, and A.H. Mehanni. 1984. Identification of dispersive behaviour and the managment of red-brown earths. *Aust. J. Soil Res.* 22:413-431.

Richards, D. and R.N. Rowe. 1977. Effects of root restriction, root pruning and 6-benzylaminopurine on the growth of peach seedlings. *Ann. Bot. (London).* 41:729-740.

Ross, P.J., S.W. Bailey, K.R.J. Smettem, and K.L. Bristow. 1992. Modelling soil water and solute movement using SWIM – Version 2. In: National Soils Conference, p. 10. Aust. Soc. Soil Sci. Inc., Glen Osmond, S.A.

Russell, E.W. 1973. *Soil conditions and plant growth.* 10th. Edn. Longmans, London.

Russell, R.S. 1977. *Plant root systems: their function and interaction with the soil.* McGraw-Hill, London.

Shanmuganathan, R.T., and J.M.Oades. 1982. Effect of dispersible clay on the physical properties of the B horizon of a red-brown earth. *Aust. J. Soil Res.* 20:315-324.

Shanmuganathan, R.T. and J.M.Oades. 1983. Modification of soil physical properties by addition of calcium compounds. *Aust. J. Soil Res. 21*:285-302.

Sharpley, A.N., J.J. Meisinger, J.F. Power, and D.L. Suarez. 1992. Root extraction of nutrients associated with long-term soil management. In: J.L. Hatfield and B.A. Stewart (eds.) *Adv. Soil Sci.* 19:151-217.

Shaw, B.T. (ed.) 1952. *Soil physical conditions and plant growth.* Academic Press, NY.

Skene, J.K.M. and T.J. Poutsma. 1962. Soils and land use in part of the Goulburn Valley, Victoria. Tech. Bulletin, No.14. Department of Agriculture, Melbourne, Australia.

Spoor, G. 1979. Effective subsoiling. In: Conference on Soil Management for Better Cereal Yields. Min. Agric. Fish. Food, Chelmsford, Essex, U.K.

Spoor, G. and R.K. Fry. 1983. Soil disturbance generated by deep-working low rake angle narrow tines. *J. Agric. Engng. Res.* 28:217-234.

Spoor, G. and R.J. Godwin. 1978. An experimental investigation into the deep loosening of soil by rigid tines. *J. Agric. Engng. Res.* 23:243-258.

Spoor, G. and R.J. Godwin. 1979. Soil deformation and shear strength characteristics of some clay soils at different moisture contents. *J. Soil Sci.* 30:483-498.

Stafford, J.V. 1979. The performance of a rigid tine in relation to soil properties and speed. *J. Agric. Engng. Res.* 24:41-56.

Stafford, J.V. 1981. The application of critical state soil mechanics: the performance of rigid tines. *J. Agric. Engng. Res.* 26:387-401.

Taylor, A.J., and K.A. Olsson. 1987. Effect of gypsum and deep ripping on lucerne (*Medicago sativa L.*) yields on a red-brown earth under flood and spray irrigation. *Aust. J. Exp. Agric.* 27:841-849.

Taylor, H.M. and L.F. Ratliff. 1969. Root elongation rates of cotton and peanuts as a function of soil strength and soil water content. *Soil Sci.* 108:113-119.

Taylor, H.M., W.R. Jordan and T.R. Sinclair (eds.) 1983. *Limitations to Efficient Water Use in Crop Production.* Amer. Soc. Agron., CSSA, Soil Sci. Soc. Amer., Madison, WI.

Taylor, H.M., G.M. Roberson, and J.J. Parker, Jr. 1966. Soil strength-root penetration relations for medium to coarse-textured soil materials. *Soil Sci.* 102:18-22.

Tisdall, J.M. and H.H. Adem. 1988. An example of custom prescribed tillage in south-eastern Australia. *J. Agric. Engng. Res.* 40:23-32.

Tisdall, J.M. and A.S. Hodgson. 1990. Ridge tillage in Australia. A review. *Soil Tillage Res.* 18:127-144.

Tisdall, J.M. and J.M. Oades. 1979. Stabilization of soil aggregates by the root systems of ryegrass. *Aust. J. Soil Res.* 17:429-441.

Tisdall, J.M. and J.M. Oades. 1982. Organic matter and water-stable aggregates in soils. *J.Soil Sci.* 33:141-163.

Tisdall, J.M., K.A. Olsson, and P. Willoughby. 1984. Soil structural management and production in a non-cultivated peach orchard. *Soil Tillage Res.* 4:165-174.

Trouse, A.C. Jr. and R.P. Humbert. 1959. Deep tillage in Hawaii: I. Subsoiling. *Soil Science.* 88:150-158.

Unger, P.W., H.V. Eck, and J.T. Musick. 1981. Alleviating plant water stress. p. 61-96. In: G.F. Arkin and H.M. Taylor (eds.) *Modifying the root environment to reduce crop stress.* Amer. Soc. Agr. Eng., Michigan, WI.

van Ouwerkerk, C. and P.A.C. Raats. 1986. Experiences with deep tillage in the Netherlands. *Soil Tillage Res.* 7:273-283.

Veihmeyer, F.J. and A.H. Hendrickson. 1938. Soil moisture as an indication of root distribution in deciduous orchards. *Plant Physiol.* 13:169-177.

Wagenet, R.J. and J.L. Hutson. 1989. LEACHM: A process based model of water and solute movement, transformations, plant uptake and chemical reactions in the unsaturated zone. Centre for Environmental Research, Cornell University, Ithaca, N.Y.

Wells, C.B. 1956. The practice of subsoiling – a brief review *J. Aust. Inst. Agr. Sci.* 22:247-251.

Wesseling, J. 1974. Crop growth in wet soils. p. 7-37. In: J.van Schilfgaarde (ed.) *Drainage for agriculture.* Monograph 17. Am. Soc. Agron., Madison, WI.

Willatt, S.T. and K.A. Olsson. 1982. Root distribution and water uptake by irrigated soybeans on a duplex soil. *Aust. J. Soil Res.* 20:139-146.

Williams, J. 1983. Soil hydrology. p. 507-530. In: *Soils: an Australian viewpoint.* CSIRO Melbourne and Academic Press, London.

Application of Mole Drainage in the Solution of Subsoil Management Problems

Gordon Spoor

I. Introduction

There are numerous physical and chemical subsoil problems limiting crop production on heavier soils, which are either directly or indirectly related to low

1-56670-020-5/95/$0.00+.50

soil hydraulic conductivity and inadequate subsurface drainage. These problems include salinity, sodicity, low fertility, inadequate root development, aeration stress and poor soil structure. With restricted water movement, it is often impossible to achieve or maintain adequate water flow rates and throughputs through the profiles, and it is extremely difficult to introduce amending materials or remove unwanted products. Any operation which will open up the profile and speed up unwanted water removal will not only increase the feasibility of satisfactory reclamation or improvements, but will also often reduce the risk of the problem developing in the first place.

Improvement of subsoil problems in this way involves the creation of a continuous network of larger pores to aid water flow and if natural drainage at depth is inadequate, the provision of some form of subsurface drainage. This improvement has to be achieved at an economic cost which can be difficult when hydraulic conductivities are low. Soil loosening techniques of various forms are available for fissure development (Spoor and Godwin, 1978), but care is needed to ensure the soil disturbance is of the right form and not excessive; otherwise the life of any improvement is likely to be short due to recompaction. Traditional pipe or open channel subsurface drainage systems are often too expensive or impractical if installed at the close drain spacings needed. In addition, in many soil improvement and reclamation situations, the drainage intensity needed is often greater at the start of the operation than afterwards. Meeting this initial need with a permanent drainage system results in an unnecessarily intense and costly system for later; yet without it either no or an inadequate rate of improvement would occur. Lower cost surface drainage systems are rarely capable of overcoming subsoil problems alone, but they can often help considerably in reducing the magnitude of the subsurface drainage requirement (Schwab et al., 1974).

One soil water management system which has evolved successfully with some development over the past few hundred years in a number of low hydraulic conductivity heavy soil areas is mole drainage (Nicholson, 1942; Hudson et al., 1962; Bowler, 1980). The advantage of this system is its ability to both modify the hydraulic conductivity of the profile and to provide a subsurface drain at much lower cost than other alternatives. The success of the system is dependent upon a range of factors including soil type and moisture condition, nature of the water problem, implement design, field installation technique and subsequent climatic and management conditions. It is not surprising, therefore, that the direct transfer of such a system to other soil areas, without modification to equipment or field technique, has not always proved satisfactory. Mole drainage is not adaptable or suitable for all heavy soils and their problems. Experience shows however, that with systematic experimentation and development, its application and use can be successfully extended well beyond the current traditional long established moling areas and situations (Spoor and Leeds-Harrison, 1986). This chapter explores subsoil situations where the mole drainage technique with appropriate modification and development has potential, or is certainly worthwhile investigating further, for alleviating subsoil problems with a minimum of profile manipulation. Guidelines are given on specific field requirements in each situation, on possible promising techniques and implement design changes, and on a systematic experimental approach which should be helpful in developing a successful system.

II. Current Mole Drainage Practices

Before exploring new mole drainage applications it is useful to clarify the drainage, soil and climatic conditions where the technique is currently used successfully, so that differences in requirements for the new applications can be identified and addressed.

A mole drain is effectively an unlined soil pipe which transmits water in a similar manner to any other form of pipe drain. Its stability is due to soil cohesion and hence it is only practical in clayey soils with strong cohesive properties. Mole drainage is currently used primarily for the control of perched water tables in clayey soils in humid areas, where the requirement is for the rapid removal of excess water from the surface soil layers. The soils usually have clay contents in excess of 30% and smectitic and/or micaceous clay minerals tend to predominate. A critical soil property having a major influence on the life of the systems is the stability of the soil structure when wetted (Childs, 1942a). Mole drains in soils with poor structural stability on wetting tend to have a shorter life. They are not used currently on dispersive soils. The required mole drain life desired by many farmers is 3-5 years or more, but this is very dependent upon the benefit/cost ratio of the moling operation.

A typical mole drainage installation also illustrating the essential elements of the mole plow is shown schematically in Figure 1. A circular drainage channel is initially formed at depth by the mole plow foot and enlarged in diameter by the following expander. The mole plow leg opens up a leg slot and generates a regular series of vertical leg fissures in the adjacent soil mass. These fissures, projecting forwards in the direction of travel at approximately 45°, together with the leg slot, effectively produce a continuous macro-pore connection between the mole channel and the surface soil layers (Godwin et al., 1981). They constitute the major water flow paths into the drain, as illustrated in Figure 2 and identified by Leeds-Harrison et al. (1982, 1986). Once saturated conditions develop on the perching layers, water flow through the connecting fissures tends to be rapid, producing a high peak flow as shown in the drain discharge hydrograph, Figure 3. With well-defined water inputs at the soil surface, drain discharge falls off rapidly after the peak and can drop to zero after 1-2 days. This discharge characteristic indicates the mole drain itself tends to carry water for only relatively short periods, flows being zero or very low between the storm inputs. The amount of water flow to the drain through the remainder of the soil profile depends upon the magnitude of its hydraulic conductivity properties and can be almost zero in unfissured poorly structured soil.

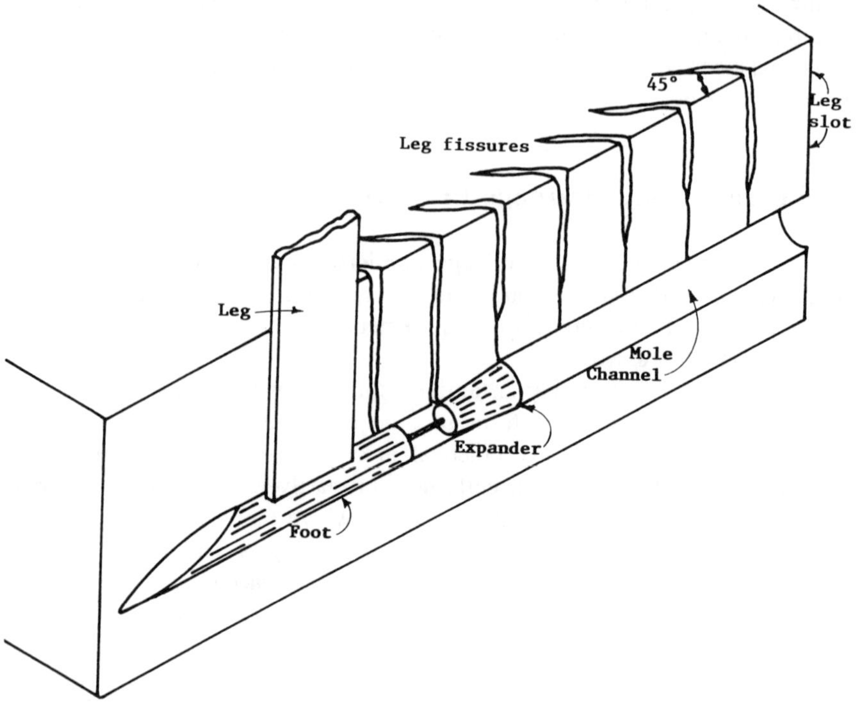

Figure 1. Schematic diagram of mole plow and resulting soil disturbance.

To achieve the required type of soil disturbance shown in Figure 1, the mole plow foot and expander are operated at a sufficient depth, to ensure that the downward soil confining stress at working depth is adequate to produce a compressive channel forming type of soil failure. Shallower working with lower confining stresses allows an upward brittle loosening failure without stable channel formation, see Figure 4. The transition working depth separating these two types of failure is defined as the critical depth (Spoor and Godwin, 1978). Successful mole channel installation can, therefore, only be achieved when mole plow working depth is below its critical depth.

No positive control is usually exerted over the grade of the mole channel at the time of installation. Nevertheless, although grade variation can be tolerated, reverse grades are whenever possible avoided, to prevent water ponding in the channel. Water ponding for long periods significantly reduces channel life. The average grade of the mole channel along its length is similar to the average grade of the field surface. Mole plow design and hitching arrangements are responsible for eliminating or reducing the effect of any small local field surface undulations on mole channel grade (Spoor et al., 1987).

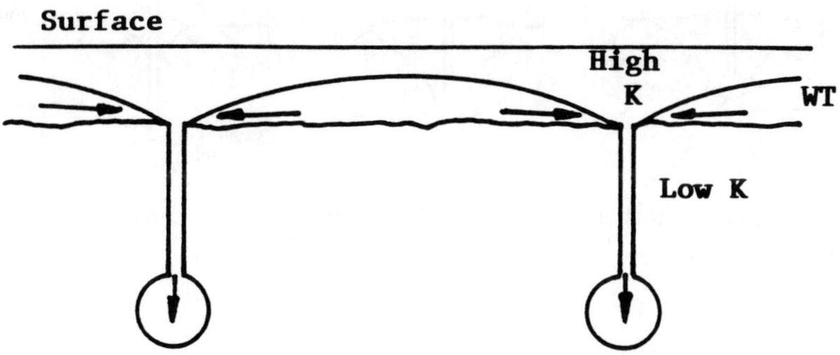

Figure 2. Water flow path in perched water table situation.

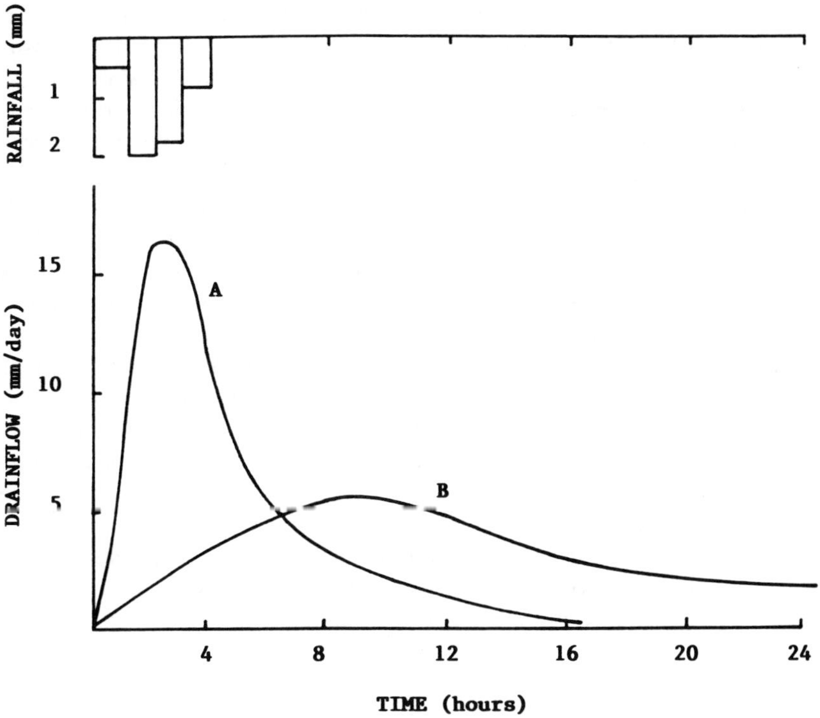

Figure 3. Typical drainflow response from a mole drain with leg fissures (A) and without leg fissures (B) in a well structured soil.

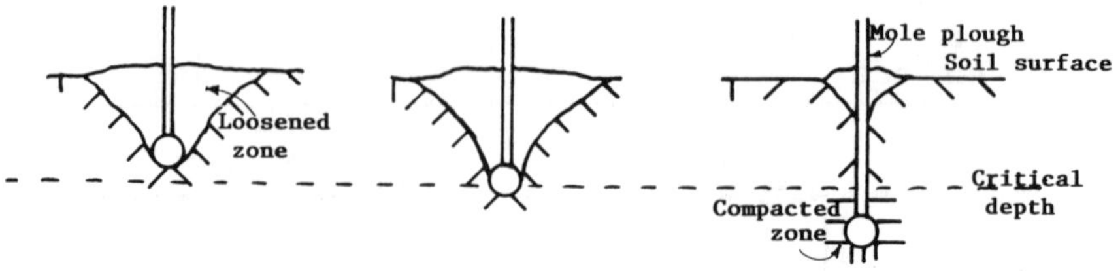

Figure 4. Soil disturbance with mole plow working above and below critical depth.

Both the leg fissures and the mole channel deteriorate with time and this deterioration has a direct influence on drainage system performance. Leg fissure closure is usually due to soil swelling or direct soil structure collapse on wetting. The closure rate depends upon the opportunity time for swelling, i.e. the time when free water is present, the swelling potential of the soil and the extent of shrinkage and hence fissure re-opening during dry periods. The latter can rejuvenate the life of cracks that have been effectively closed. As crack closure occurs, water transfer rates to the drain decrease and the drain discharge hydrograph becomes much less peaky, with a long tail recession period during which the drain continues to discharge water at a very low rate, see Figure 3 (Leeds-Harrison et al., 1982).

The manner and rate of mole channel collapse is very dependent upon the installation and subsequent wetting and drying conditions, and the soil's swelling and shrinkage properties. Life may vary from 10 minutes to 10 years or more, and deterioration and failure usually occur through one or more of the following failure mechanisms as identified by Spoor and Ford (1987):

a) Roof collapse or "expander failure": roof fall occurs due to inadequate cohesion between the soil structural units or between failure planes created by the expander in the roof area. The roof shape tends to take on a pointed arch form in the latter case, see Figure 5a. This type of failure is common following installation under rather drier conditions or when packing in the roof area has been poor due to low confining stress, e.g., weak soil above. It is also common in soils with very stable micro-aggregates which do not bond tightly together when compressed. Larger diameter channels are at greater risk from this type of failure than smaller ones.
b) Unconfined swelling: here, mole channel diameter decreases with time as a result of soil swelling into the channel void, see Figure 5b. This is common under conditions of prolonged wetting in the absence of sufficient drying to induce shrinkage and recovery in channel size. The larger the initial channel diameter the longer its life.
c) Cyclical swell/shrink failure: repeated wetting and drying cycles causing alternate swelling and shrinkage weakens the channel roof and wall areas inducing eventual collapse, see Figure 5c. Excessive shrinkage during

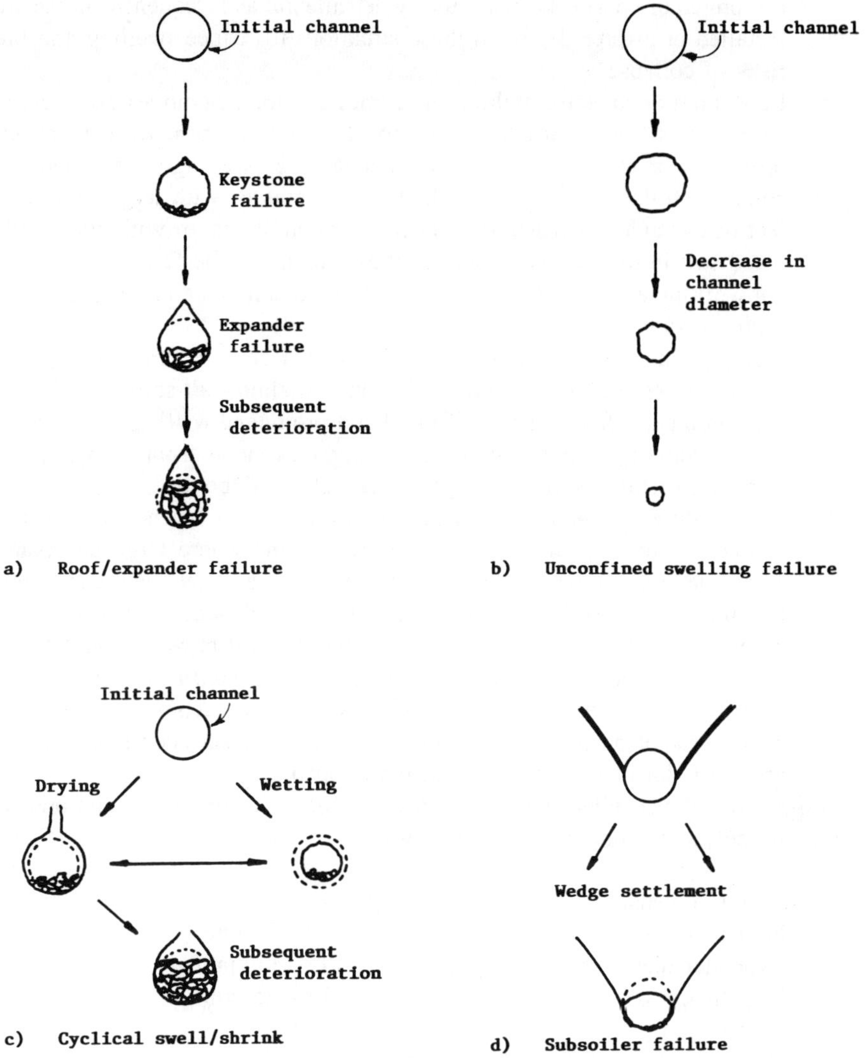

Figure 5. Types of mole channel failure.

prolonged dry periods is particularly detrimental and frequently moles are installed at greater depths in these situations to reduce swelling and the risks of collapse.

d) Loosening or subsoiler failure: the complete roof area moves downwards en masse as a result of a loosening soil failure at the time of installation, producing a channel of flattened appearance, see Figure 5d. This is common following moling at shallower depths or with larger diameter feet or expanders, particularly under drier conditions, or with unconsolidated soil in the surface layers. In these situations the feet or expanders are working either at or above their critical depth and a brittle loosening failure results.

e) Slurry failure: the channel fills with slurried soil. A common type of failure on less stable structured soils, and on almost all soils if moled in the presence of free water, or if rapid and prolonged wetting occurs soon after installation. Some of the slurrying may arise from dispersed or weakly structured soil entering the channel from above.

f) Infill failure: channel filling with soil from surface layers. This occurs following surface cultivations under conditions where large shrinkage cracks have developed in the leg slot and fissure areas above the channel.

g) Erosion failure: soil is eroded from the channel largely as a result of higher water flow rates. This type of failure is most prevalent on the less stable structured soils, in those suffering from unconfined swelling and subsoiler failure, and in channels with excessively steep grades. Leg slot erosion can also sometimes be severe, due to rapid surface runoff into an open leg slot or fissures in an unstable topsoil.

Mole plow dimensions and design features of mole drain installations in common use can be summarized as follows:

Mole plow dimensions:	
Foot diameter	50 - 75 mm
Expander diameter	75 - 100 mm
Leg thickness	15 - 25 mm
Side length of leg	200 mm
Installation features of mole drain layouts:	
Channel depth	0.4 - 0.7 m
Spacing	2.0 - 3.0 m
Length of drain run	20 - 100 m
Channel gradient	1 - 3%
Channel life	3 - 15 years
Soil conditions:	
Clay content	30% +
Texture uniformity	good
Aggregate stability on wetting	moderate to good
Moisture status at moling depth	lower section of plastic range

While differences do occur between mole drainage installations, the above descriptions and failure types are representative of the majority of current situations. They also provide a conceptual framework to identify where changes may be needed to satisfy new applications, as well as improving the life and performance of existing ones.

III. Current Equipment and Installation Considerations

There is a range of different types of field equipment and techniques currently in use for mole drainage, selection being dependent upon site conditions and power availability. Full details are provided in Nicholson (1942); Hudson et al. (1962); and Bowler (1980).

A. Equipment

The major differences between mole plows in current use, lie in their grade control mechanisms and hitching arrangements, rather than in their foot, leg and expander assemblies which tend to be very similar. Mole channel grade is almost always controlled by the mole plow/power unit combination itself, without any assistance from external systems such as rotating lasers. This "self" grade control is very feasible due to the considerable stabilizing influence of the mole plow foot (Childs, 1942b). The grading capabilities of the different mole plow types differ markedly and their suitability is very dependent upon the extent of local field surface undulations, relative to the average gradient of the field surface. In most mole drainage installations the average gradient of the mole channel is similar to that of the field surface. A well designed mole plow should, wherever possible, isolate the mole plow foot and hence channel from any perturbations caused by local surface undulations. In particular it is necessary to avoid reverse grades in the channel which induce water ponding, and sudden grade changes which accelerate roof collapse and can cause subsoiler failure.

The four basic types of mole plow in common use, illustrated schematically in Figure 6, are:

a) fully mounted, where the plow is attached directly to the tractor 3-point linkage. Depth is controlled by adjusting linkage geometry and leg length.

b) long beam scrubbing mole plow. Here a long beam (3-4 m long) carrying the mole plow leg and foot slides along in contact with the soil surface. Depth is adjusted by changing leg length.

c) long beam mole plow with front skids. The beam runs clear of the ground surface, except where it is supported on skids at the front end. Depth is controlled by angling the leg and foot assembly relative to the beam and changing leg length.

d) long beam floating mole plow, where the complete beam rides clear of the soil surface (Spoor et al., 1987). This plow is usually attached to the tractor

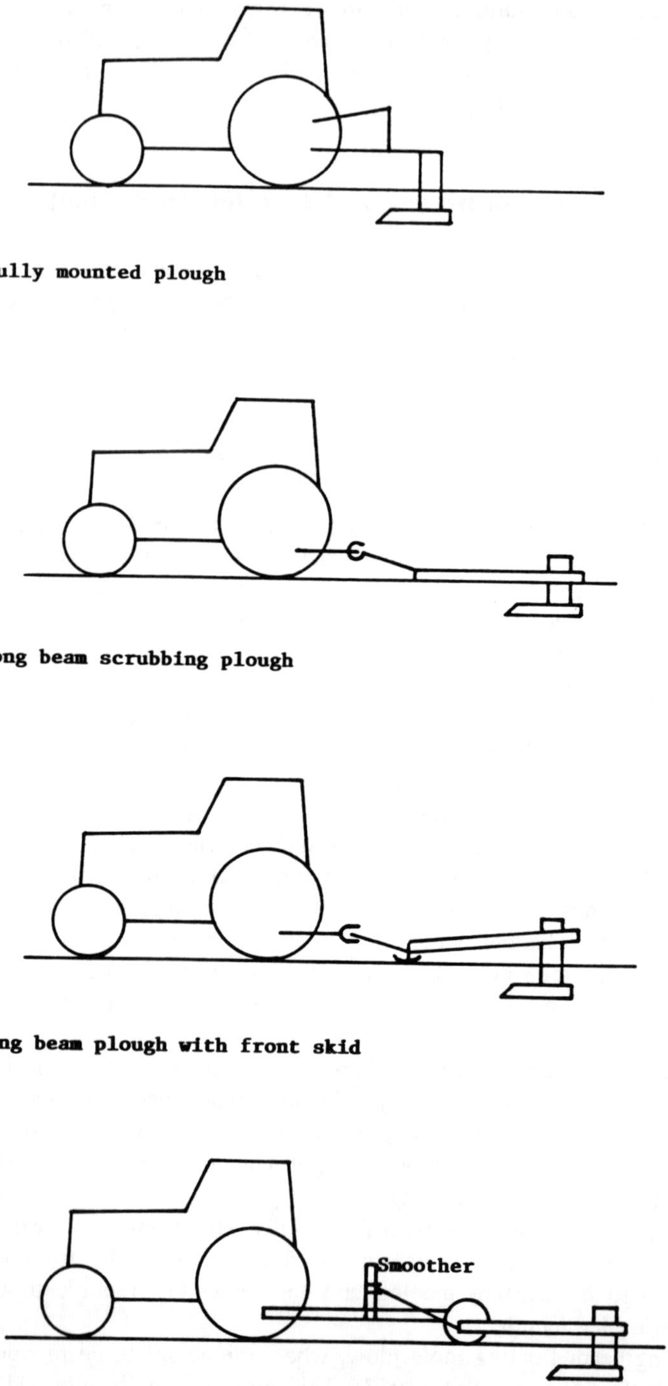

Figure 6. Types of mole plow.

through a smoother device, to reduce the effects of tractor pitching on mole foot movement on undulating surfaces. The foot operates at the required depth, parallel to the average grade of the soil surface. Deviations at the mole foot are minimized by hitching the plow to the center of the smoother and locating the smoother/tractor hitch beneath the rear axle or the center of the tracks, on wheeled and tracklaying tractors, respectively.

The fully mounted mole plow is the least satisfactory of all the types, unless it is operating on graded surfaces or on surfaces where local undulations are very minor. Best results are achieved when it is operated with a "free" tractor linkage rather than in draft control, but even then, channel grade is severely affected by tractor pitching. It is also unsuitable for moling from open ditches due to the close proximity of the leg and foot assembly to the tractor. This type of plow is only recommended for moling on graded surfaces, when it must be operated with the tractor linkage in "free float". Some close hitched mole plows are depth wheel or skid controlled and again these are only satisfactory on graded surfaces.

The long beam scrubbing and front skid mole plows are more satisfactory than the mounted type when local surface undulations are more significant. The scrubbing long beam tends to bridge over the irregularities minimising movements at the foot, particularly when passing through local depressions, but it can suffer from sudden pitching as it rides over surface rises. This scrubbing plow suffers, however, from having a relatively higher draft, due to the sliding forces generated at the beam/soil interface. In work, the forward tilt of the mole foot is adjusted so that the downward force generated is just sufficient to keep the beam in contact with the surface. Any further tilt simply increases draught without improving performance and the increase can be as high as 100% with a poorly adjusted plow, (Godwin et al., 1981). The additional draft penalty is less with the front skid plow, which again is adjusted so that the skids are just in contact with the soil surface. Angular movements and hence grade changes due to the skid rising and falling over undulations are minimized through having a long beam and relatively short foot.

Both the scrubbing beam and front skid plows are used for moling from open ditches, but care is needed to avoid reverse grades developing in the mole channel at the critical outfall position. This risk is greatest when the ditch berm is higher than the field surface. The scrubbing beam can only be used in these circumstances if the additional berm height is bulldozed away before moling. The front skid type will function satisfactorily only if the skid is at field surface level ahead of the berm; usually a rare occurrence. These two types of plow are currently in most common use.

The long beam floating plow always operates with minimum draft and the foot position and hence the mole channel itself is least affected by surface irregularities and soil variations. This plow also allows the mole channel to be graded on level surfaces without the need for sophisticated grading equipment. Grading is achieved by raising or lowering the height of the mole plow hitch at the smoother while moling is in progress.

a. Long beam scrubbing mole plough

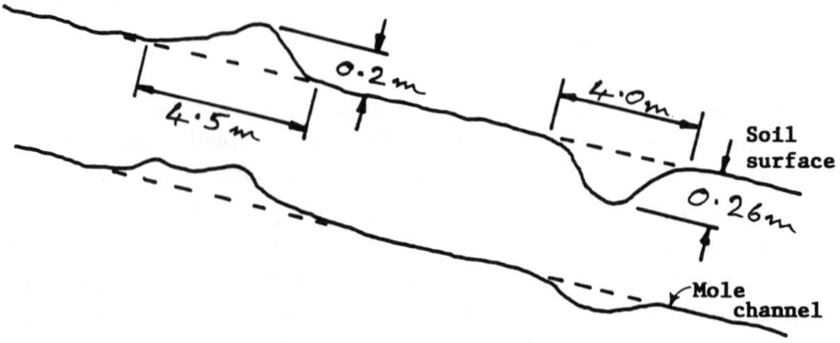

b. Long beam floating mole plough

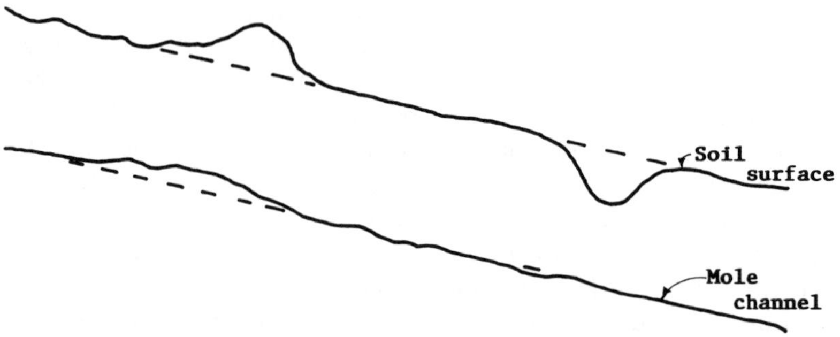

Figure 7. Mole channel grade variation with long beam scrubbing and floating mole plows (NOTE: exaggerated vertical scale).

The improved performance of the floating as compared with the scrubbing plow, in terms of reduced grade variation when moving across surface undulations, is clearly shown in Figure 7. Of particular significance also from a channel stability viewpoint is that when grade changes do occur with the floating plow, they are gradual rather than sudden. While it is possible to operate both the long beam scrubbing and front skid plows in a floating position, in the absence of a smoother device they both suffer more severely the adverse effects of tractor pitching.

The long beam floating plow is ideally suited to moling from open ditches, providing the rear wheels on the smoother do not come into serious contact with a higher ditch berm. The floating plow is also less likely to suffer blockage when moling in the presence of surface trash than the scrubbing plows, which can lift as trash builds up beneath their beams.

B. Installation Considerations

The success of any mole drainage installation in a given soil is very dependent on due consideration being given to and action taken on all the critical factors likely to control channel stability. These factors, equally critical for any future extension of mole drainage practices are now considered.

1. Timing

Unless soil moisture conditions at moling depth are in the plastic range, it will not be possible to achieve the firm, smooth-walled channel desired for stability. Channel walls and the channel shoulder area become much rougher under drier moling conditions and more susceptible to collapse.

Long term channel stability on many soils is dependent upon an aging period without significant wetting, after formation. This period enables soil structural bonds broken during the moling operation to reforming and increasing structural stability and resistance to swelling (Spoor et al., 1982). Installation should, therefore, be timed whenever possible to precede a dry period of ideally 2-3 weeks at least, to allow this aging process to proceed before the channel is stressed with water. Moling in the presence of free water in the profile is frequently disastrous.

2. Depth

In the perched water table situations where mole drainage is currently practiced, once the drain is in the perching horizon, there is no drainage benefit from draining deeper. Depth selection is, therefore, governed by other considerations which include:

a) the most stable soil horizon for channel formation
b) depth of significant shrinkage crack development during dry periods
c) magnitude of surface loads
d) depth of outfall available
e) size of power unit available.

Installation in the most stable horizon below depths of significant shrinkage is ideal if achievable. Depths of 0.4 m and 0.5 m are often considered the minimum in grassland and arable situations, respectively, to avoid damage from surface loads. Soil freezing has not been found to have a significant effect on channel stability and hence is not an important depth selection factor (Eriksson et al., 1987; Weil et al., 1991). The selected depth must match the traction available, to allow a consistent steady pull with minimum wheel or track slip.

Once an appropriate depth has been decided upon, foot and expander size are then selected to ensure a "below" critical depth type of disturbance.

3. Spacing

Due to the low cost of mole drainage, spacing is not usually an important consideration. Spacing is usually reduced as the risks of mole channel collapse increase. Spacings of 2.0-2.5 m are common.

4. Length of Run

Like spacing, run length is dependent upon collapse risk, but it is also influenced by topography, the risks of reverse grades developing and the likelihood of erosion. The greater the risk of problems, the shorter the length of run.

5. Gradient

The gradient selected is largely dependent on the minimum required to avoid reverse grades, the maximum to avoid erosion and the need where necessary to intercept surface water. Common gradients range between 0.5-3%.

6. Outfalls

In common with other types of drainage system, mole drain performance can only be as good as the outfalls and drain connections allow. Three broad types of outfall are used, namely, open ditches or sumps, discharge into permeable backfill overlying a permanent pipe system, see Figure 8, and discharge into a deeper mole drain, often termed a moled main or major mole.

Open ditch outfalls are satisfactory providing each mole channel exit is protected by a permanent pipe insert, there is no reverse grade at the outfall and the ditch is well maintained. Problems of channel wall shrinkage local to the outfall, causing collapse during dry periods in smectitic clay soils, can be minimized, by extending the length of the permanent pipe insert within the mole channel itself.

To minimize outfall maintenance and maximize mole channel protection, the more expensive permeable backfill pipe outfalls are now in more common use, see Figure 8. Water moves downwards from the mole channel through the permeable backfill into the pipe for discharge. For rapid water discharge the backfill must be stable, permeable and free from fines less than 5 mm diameter

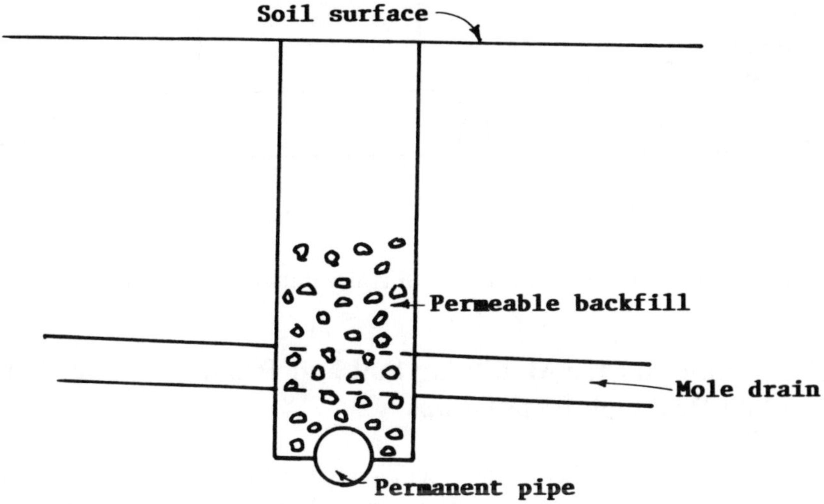

Figure 8. Mole channel/permeable backfill connection.

(Dennis and Croote, 1977). Common backfill materials are gravel or crushed stone in the size range 20-40 mm diameter.

Narrow bands of permeable backfill (100-150 mm width) are attractive to reduce the quantity of expensive material required. There is a risk, however, under wetter conditions and with newly installed backfill, that the mole plow will draw soil into the backfill, sealing off the connection (Mulqueen, 1971; Castle et al., 1992). Attaching a narrow tine behind the expander to work approximately 10 mm deeper, see Figure 9, overcomes this problem by disrupting the soil seal, hence remaking the connection.

The moled main system was the original outfall system, used for many years and is extremely cheap (Nicholson, 1942; Hudson et al., 1962). One, two, or three deeper mole channels are installed first (1-2 m apart) to act as collector drains, and the field or minor moles pulled across them. The invert of the field moles is usually 25-50 mm above the top of the moled main. In well structured soils, the leg slot of the moled main or natural structural fissures provide the connection for water discharge from the field mole into the main. Under wetter conditions in poorly structured soils, a metal spear 10-15 mm diameter is forced between the two channels to make the connection (Spoor et al., 1992). The moled mains discharge into an open ditch or pipe through a protected outfall for stability. This system functions well on soils of good moling quality. Failure risks are high in marginal moling situations and hence it is only appropriate to use moled mains in these cases when a relatively short life is acceptable.

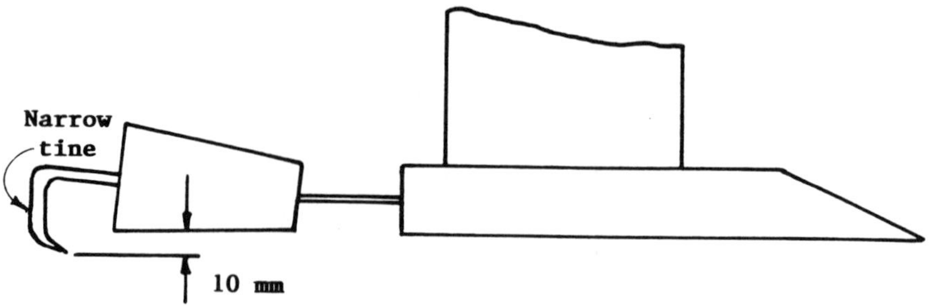

Figure 9. Expander tine to disrupt soil seal within permeable backfill.

IV. Techniques for Modifying Soil and Drainage Conditions in Mole Drainage Installations

The soil, drainage and installation conditions described in the previous sections have proved satisfactory for perched watertable control in humid areas with relatively intermittent inputs of storm water. The equipment used for the installations has largely evolved with time, to provide the required conditions for drainage on the local soil types, under the prevailing soil and climatic conditions. Where similar equipment has been used for mole drain installation under different sets of conditions, the results have often proved disappointing or unsatisfactory. The reasons for this lack of success have largely been attributed to soil unsuitability. Usually in these trials however, little attempt was made to modify the equipment or technique, to determine whether any such changes could modify soil conditions to help overcome the problems experienced. For the successful extension of the moling technique into new situations it will, therefore, be often necessary to modify the soil conditions produced at installation to meet the new requirements. Early work reported in Nicholson (1942) and Hudson et al. (1962) and further detailed studies in Silsoe College, U.K., have clearly demonstrated that implement design and field installation technique changes can significantly modify the soil conditions produced at mole installation. Providing the water flow regime conditions required and potential soil and climatic limitations in any new situation can be identified, implement and technique modifications open up possibilities for successfully extending the moling technique to new situations. Before considering the potential new applications for mole drainage, possible ways of modifying soil and drainage conditions through changes to the equipment and installation technique will be discussed. The modifications are considered within three groupings:

a) Changes to the water flow regime to moderate both flow rates and routes, largely by modifying leg fissure development and fissure continuity.

b) Changes to increase channel stability by countering specific channel failure mechanisms.
c) Changes to increase the inherent soil structural stability.

A. Modifications to Water Flow Regime

Modifications to the water flow regime can be achieved by manipulating leg slot size, the size and frequency of the leg fissures and the continuity of the leg slot and fissures between the soil surface and the mole channel.

The leg slot width soon after installation is a direct function of leg thickness and the elastic recovery properties of the soil. The latter increase with increasing plasticity. Changing leg thickness and selecting appropriate moisture conditions for installation can, therefore, control slot width. The thinner the leg and the more plastic the soil condition, the greater the chance of achieving a very narrow or closed leg slot. Leg fissure development is dependent upon three main factors: the magnitude of the sliding resistance between the soil and the side of the mole plow leg relative to soil shear strength, soil air filled porosity and existing fissuring or cracking in the soil mass (Hopewell, 1953; Spoor et al., 1989). Saturated soils with little previous crack development are extremely difficult to fissure at the time of moling. Nevertheless, planes of weakness can be established under these conditions which will open up in shrinking soils, if a drying phase can follow installation. Extremely large vertical fissures can be created in drier cracked soils.

Under given soil moisture conditions, the size of and spacing between leg fissures can be modified within limits, by changes to leg thickness and/or leg side area. These leg geometry modifications change soil/metal sliding resistance at the side of the leg and hence fissure development. Increasing the sliding resistance tends to create a smaller number of wider, more widely spaced leg fissures and vice versa. The most appropriate leg modifications to use to adjust fissure size depends upon the relative magnitudes of the soil/metal friction and adhesion components of the soil/metal sliding resistance. Under very sticky conditions when adhesion is high, fissure size is most sensitive to leg side area. When friction dominates, leg thickness and side roughness are the most important factors. Fissure size can, therefore, be increased by increasing side area, or leg thickness or both, depending upon soil condition. Figure 10b, taken from Spoor et al. (1989), shows the effect of changing leg thickness on fissure spacing, width and depth (defined in Figure 10a) (Spoor et al., 1989). In these tests soil moisture content was just above the plastic limit, adhesion was low and the air filled porosity was 9%. Crack depth, measured perpendicular to the leg slot, see Figure 10a, always decreases as the mole channel is approached.

Crack size also increases with leg surface roughness, but due to continuing leg wear reducing roughness with time it is rarely a practical variable. The extent to which leg thickness can be increased or decreased is limited by mole foot diameter and mechanical strength factors respectively. Too thick a leg relative

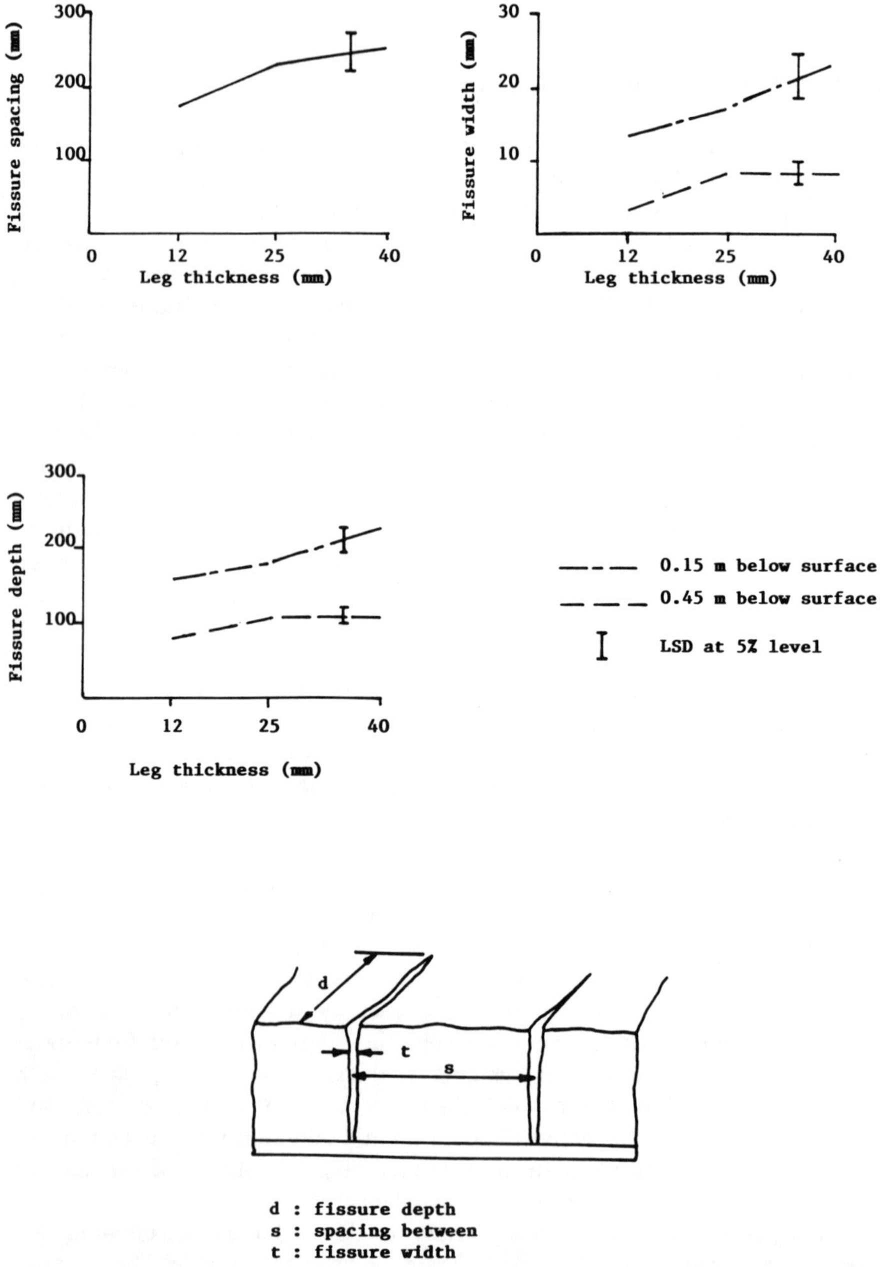

Figure 10. Leg fissure geometry.

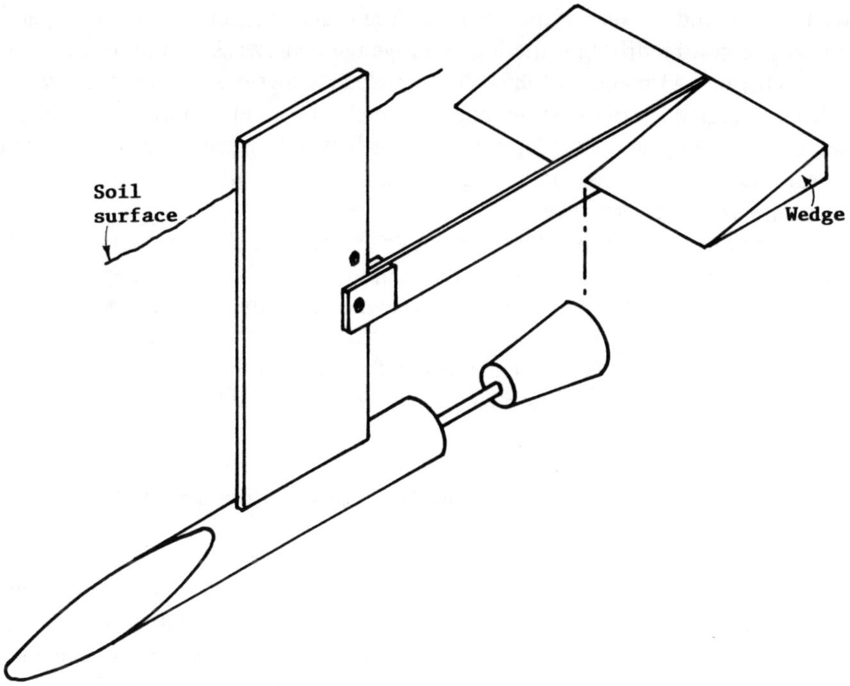

Figure 11. Trailing wedge to disrupt leg fissures and slot.

to channel diameter, leaving a wide leg slot, produces a very weak channel roof. Increasing side area is a useful alternative to changing leg thickness in such situations. The continuity of the water flow path from the soil surface layers to the mole channel through the leg fissures and slot can be throttled or disrupted by manipulating fissure and slot size and sealing off connections. Reducing leg slot and fissure thickness will assist in reducing flow rates, and crack closure can be most readily and conveniently achieved in the surface layers and at the channel/leg slot connection.

Leg slot and fissure closure in the surface layers can be achieved through direct compaction. Due, however, to the resistance of vertical cracks to closure under vertical loading, it is necessary to disrupt them first before compaction. This disruption and loosening can be effectively achieved using the wedge device shown in Figure 11 (Gebre-Michael, 1989). Following lifting and fissure disruption with the wedge, the soil can be recompacted immediately with tractor wheel or heavy roll to destroy all direct fissure connections with the surface. Care must be taken to ensure the channel below is not damaged during this recompaction operation. Crack sealing is most efficient with soil in a plastic condition at wedge working depth. It is essential that the leading edge of the wedge is behind the rear of the expander (see Figure 11), otherwise the expander will move soil readily into the loosened wedge zone weakening the roof of the mole channel. Ormandy (1985) developed an experimental mole plow with sideways angled leg to assist slot and fissure closure and to improve channel roof stability. Loading above the leg slot with this installation produces effective crack closure. Development is continuing at Silsoe College with this plow to improve its commercial performance.

Sealing the connection between the mole channel and the leg slot is achievable by increasing expander size relative to mole plow foot size. The larger the expander diameter relative to foot diameter the better the sealing, up to the limit where the expander induces an upward "above" critical depth loosening type of soil failure. The narrower the leg slot and fissures and the more plastic the soil, the more easily this connection can be sealed or throttled.

B. Modifications to Counteract Specific Types of Channel Failure

As previously discussed, the mole channel can be subjected to a range of failure mechanisms each inducing a different type of failure. The prevailing failure mechanisms differ between moling situations and the installation technique adopted should reflect these differences to form a channel most capable of resisting the particular failure stresses likely to be imposed. Experimental work and field experience (Nicholson, 1942; Hudson et al., 1962; Spoor & Ford, 1987; Spoor, 1987) have established that channel failure resistance to many of these mechanisms can be increased by:

a) Modifications to foot and expander shape, size and operating depth.
b) Appropriate installation timing with respect to soil moisture and weather conditions.

Possible modifications to increase channel stability through counteracting specific types of failure mechanism are discussed in the context of the failure types.

1. Roof Collapse/Expander Failure

The requirement here is to increase cohesion in the roof area between the soil structural units, particularly if fine, and between the expander disturbed soil and the surrounding "undisturbed" soil. The risk of this type of failure increases as the soil at moling depth becomes less plastic and as mole channel diameter increases. A number of approaches are possible to reduce this failure risk, summarized below:

a. Avoid the Use of an Expander or Reduce Expander Size

This approach is appropriate when the major cause of roof collapse is failure between the expander induced failure planes and the surrounding soil. After collapse through this mechanism a pointed arch-shaped roof section is left, see Figure 5a. In most of these situations, usually under drier conditions, fitting an expander just slightly larger than the foot offers a better solution than eliminating the expander completely. This slightly larger expander smooths out channel wall irregularities increasing stability in that region, without significantly displacing soil in the roof area, hence avoiding potential roof collapse.

b. Increase Expander Diameter Relative to Mole Foot Diameter

This is an alternative approach to "a", successful at higher moisture contents and when increased packing or re-working of the soil structural units is required. It is only successful, however, providing the downward confining stress exerted by the soil above moling depth is sufficient to allow a "below" critical depth moling failure. Insufficient confining stress will induce a subsoiler type failure. A similar packing result can, however, be achieved with less risk of a subsoiler type failure by reducing the foot size relative to the expander, keeping the expander size constant.

c. Reduce Channel Size

Large diameter mole channels are more susceptible to roof collapse than small ones, hence reducing foot and expander size can help reduce collapse risks.

d. Ensure the Mole Foot Runs Parallel to the Required Channel Grade

A mole foot angled downwards rather than running parallel to the desired grade forms an oval shaped channel, reducing the ability of the following expander to pack the soil tightly into the roof area. A similar oval shaped channel with reduced stability is also formed following a sudden grade change (see subsoiler failure later).

e. Use a Tapered Rather than Barrel Shaped Expander

Two types of expander shapes are in common use, the tapered type and the barrel type which has a blunt leading edge. The barrel type tends to force soil ahead, which under marginally dry moling conditions leaves a weak roof area, more susceptible to collapse. Soil movement generated by the tapered expander is more vertical, tending to produce a more compacted and stable channel roof (Ormandy, 1985; Castle et al., 1992).

f. Delay Moling until Moisture Conditions Are More Favourable at Moling Depth

Soils at or just slightly above their plastic limit are most susceptible to expander failure and hence delaying the moling operation until higher moisture contents can be achieved will reduce the failure risk. In certain dense clay soils, moisture contents even when at saturation, are still marginally dry for stable mole channel formation. In these situations approach "a", with minimum expander size, is usually the most successful.

2. Unconfined Swelling Failure

In many situations while it is impossible to avoid unconfined swelling, it is often possible to reduce swelling rate and so increase the time period between channel formation and channel void disappearance. In any soil with a given swelling capacity, the overall rate of swelling is a function of the time of free water presence and the extent of drying and hence shrinkage between wetting periods. The opportunity periods for wetting and drying can be minimized and maximized respectively, by providing a good channel outfall to allow rapid water discharge and by avoiding reverse grades in the mole channel itself. Reverse graded sections are particularly vulnerable to unconfined swelling failure due to the much longer water ponding periods experienced in those areas.

Channel roof collapse is much less likely in areas where unconfined swelling failures dominate and hence larger diameter mole channels can be confidently contemplated to prolong channel life. The larger the void, the greater the time taken to close and in certain situations unconfined swelling may cease before complete channel closure occurs. Unconfined swelling failure can be significantly delayed if not prevented, by either filling the mole channel with a porous medium or fitting it with a stabilizing liner capable of resisting the soil swelling pressures generated. One porous medium which has proved very successful (Mulqueen, 1985) is gravel or stone and encouraging possibilities exist for the use of waste products such as rice hulls (Christen 1993, personal communication) etc, in areas where these are available at an acceptable cost. Thin walled, weak, plastic mole drain liners were tested in the 1950s and 60s with some success (Schwab, 1955), although some failures did occur with larger diameter liners (Fouss, 1965). This concept is currently being re-explored in the United Kingdom (Spoor et al., 1992) for smaller diameter liners. These measures also reduce the failure risk through other failure mechanisms.

3. Cyclical Swell/Shrink Failure

The main approaches to reducing the risk of cyclical swell/shrink failure are the control of channel size and installation depth. Smaller diameter channels offer more resistance than larger ones to the stresses imposed by alternate swelling and shrinkage, but the main defence mechanism is installation depth. The extent of swelling and shrinkage reduces with depth and hence deeper channels are less susceptible to this type of failure than shallow.

4. Loosening/Subsoiler Failure

Subsoiler failure occurs following a mole channel installation with the plow working at or above its critical depth. This problem can be overcome by increasing moling depth or by modifying or manipulating foot and expander size, surface confining stress and soil moisture content. Reducing expander size effectively raises critical depth, increasing the chances of successful moling at the same installation depth. Mole installation in a compact undisturbed profile benefits from maximum surface confining stress, which reduces the risk of subsoiler failure. Subsoiler failure is also less likely to occur when moling at higher soil moisture contents than at low. A compact soil profile with drier and hence stronger conditions at the surface than at depth offers the greatest resistance to this type of failure.

A further cause of subsoiler failure is sudden changes in mole channel grade during installation (Hann, 1993, personal communication). Sudden grade changes are accompanied by the rapid vertically upward acceleration of the front or rear of the mole plow foot. This rapid upward movement tends to induce a loosening type of soil failure in situations where moling depth is fairly close to critical depth. In addition, an oval channel is formed which increases the risk of a roof/expander type of failure due to the inability of the expander to compact the roof area. Both these failure risks can be significantly reduced by improving the grading characteristics of the mole plow, particularly in terms of sudden grade changes.

5. Slurry Failure

The prime requirement to minimize the risk of slurry failure in any given situation is to avoid the immediate and prolonged wetting of the mole channel soon after formation. The longer the soil aging period, even without drying, before the channel becomes wetted, the greater the chance the channel will be able to withstand excessive unconfined swelling and hence avoid slurry failure. Slurry problems also arise through less stable soil moving into the channel from above. The less structurally stable or more dispersive a soil the greater the risks of a slurry failure.

This type of failure can be reduced by moling at the beginning of a dry period, so that time is available for the soil to age and restabilize after disturbance, before it is wetted. When free water is present at moling depth at the time of installation, a sacrificial moling to de-water the profile before the final moling will help minimize the problem. Closure of the leg slot and fissures at the channel will help reduce the movement of less structurally stable soil from above into the channel.

6. Infill Failure

Surface soil infill into the channel can be minimized by closing off or reducing the size of the leg slot and fissures. Where mole installation is to be followed by an extended drying/shrinkage period which will maximize fissure size, the smallest possible fissures should be produced at installation.

7. Erosion Failure

Erosion failure can be controlled by reducing the channel gradient and forming channels with a high resistance to roof collapse and subsoiler failure. Small soil flakes commonly found on the channel wall are more likely to erode in channels pulled up the slope rather than down (Nicholson, 1942; Mulqueen, 1971). The difference due to installation direction in terms of erosion is, however, usually very small in practice. Leg slot and fissure erosion can result from rapid surface water inflow into open slots and fissures. This form of erosion can be prevented by eliminating the direct connection between the surface fissures and slot and those at depth connecting with the mole channel. Operating with the wedge device behind the mole leg, as discussed in the water regime modification section, to disrupt the fissures followed by immediate soil recompaction helps avoid this erosion problem.

C. Modification of Inherent Soil Structural Stability

The long term stability and life of any mole drainage system is critically dependent on the structural stability on wetting of the soil in the mole channel itself. Mole channel soil which slakes badly or disperses on wetting is unsuited to mole drainage, unless a very short channel life is acceptable. Dispersive soil channels rapidly collapse through either unconfined swelling or slurry failure and those which slake badly, deteriorate through general wall disruption or continual roof collapse. Mole channel soil stability to offset some of these effects can be increased in three ways, namely, by:

a) allowing the disturbed soil to age and strengthen before being subjected to disruptive stresses

b) modifying the electrolyte concentration of the drainage water
c) adding soil stabilizing materials.

By allowing an aging period after installation before the soil is wetted, many of the structural bonds damaged during the moling operation reform, increasing the soil resistance to swelling and dispersion (Spoor et al., 1982). Timing an installation to precede a dry period of ideally 2 to 3 weeks at least is, therefore, particularly helpful in increasing soil stability.

Increasing the electrolyte concentration of the drainage water buffers dispersive elements such as exchangeable sodium, hence increasing structural stability. This can be achieved by either introducing an available electrolyte source into the water flow path, or by ensuring the potential drainage water picks up soluble salts already in the profile en route to the mole drain. Experiments by Muirhead et al., (1993) have shown the installation of gypsum enriched slots (Blackwell et al., 1990) above mole channels in an irrigated vertisol have significantly improved channel stability. Cronin (1982), working in a very unstable saline alkali soil in Egypt, achieved very stable mole channels by ensuring good quality, low electrolyte concentration irrigation water could not directly enter the mole channel.

The direct addition of soil stabilizing materials such as gypsum, calcium chloride or stabilizing polymers into the channel at installation offers further possibilities to improve mole channel soil stability.

V. Potential Applications for Mole Drainage in Subsoil Problem Areas

There are a number of subsoil problem areas and situations where the mole drainage technique with development could assist in alleviating current difficulties. These possibilities are discussed with comments on desirable water flow regimes and on issues related to equipment and installation requirements. These comments, together with the possible types of technique modification identified in Section IV, should be helpful in any investigational and development program. Whatever the application, success will be dependent upon achieving the desired flow regime and stable mole channels. The required life and hence the degree of long term stability of the mole channel needed can, however, vary considerably. Depending upon the application and the benefit/cost ratio, the minimum acceptable life may range from days to years.

A. Perched Water Table Control in Humid Areas

This application is similar to that in most of the current mole drainage installations, where a perched water table develops in a moderately or highly permeable topsoil which overlies a low permeability subsoil. The current use of the technique is, however, restricted to a relatively narrow range of soils, areas and climatic conditions. To extend the technique to a wider range of conditions will require in some situations modifications to the equipment and field technique to counteract the particular failure mechanisms operative, or in others, simply paying more attention to detail, to ensure maximum chances of success.

Problems with mole drainage have arisen in the past in soils of reasonable aggregate stability, due to the leg fissures either being too wide or insufficiently developed, causing channel instability or restricted water entry into the channel respectively. These faults can be overcome through modification to leg geometry and the timing of installation. Others have been related to rapid channel collapse through any one or more of the different failure mechanisms. Unconfined swelling and slurry failures have tended to be the most common in wetter areas where minimal drying occurs, and roof/expander and cyclical swell/shrink failures in the drier areas with significant drying and shrinkage. All of these problems can be counteracted to some degree by technique modification.

Experience and reviews of failed mole drainage installations and trials in areas with little or no previous moling experience suggest, however, that one of the major causes of failure is lack of attention to the fundamental requirements for success, identified in Section III. Common installation faults identified include:

a. use of mole plows comprising of a subsoiler/ripper leg and foot with a trailing ball attached, to form what could only be an extremely weak channel
b. use of depth wheel controlled or fully mounted draft controlled mole plows reacting to local surface undulations and soil shear strength changes respectively, with consequent rapid changes in mole channel grade
c. mole channels pulled from open ditches with reverse grades at the outfall, due to the ditch berm being higher than the field surface; often no pipe insert into the mole outfall to provide protection against erosion
d. moling across pipe systems with no positive permeable connection between the mole and the pipe for water discharge
e. surface loosening or deep soil ripping before moling, reducing confining stress and inducing an upward loosening type of soil disturbance.

These faults must be rectified before any judgment can be made on the mole drainage potential of the soil and before any systematic technique and equipment modifications can be made to improve channel stability.

B. Perched Water Table Control in Irrigated Areas

On numerous heavy, poorly structured, flood irrigated soils, perched water tables develop during irrigation and persist for some time afterwards. Effective shallow drainage to rapidly remove this perched water would improve root zone aeration and hence crop production in all situations (Sommerfeldt, 1984). In situations with saline groundwater, further benefits in the form of reduced surface salt loads could arise from intercepting this perched water. The rapid removal of the perched water should reduce deep seepage and hence groundwater recharge. This in turn would reduce the drainage requirement to maintain groundwater tables at acceptable levels and hence minimize the quantity of salt brought to the surface for disposal. The quality of the perched water is often much better than that of the groundwater and therefore, would be more suitable for re-use. Relatively closely spaced mole drains on suitably stable soils could provide the necessary shallow drainage.

The water flow regime required in this application is one which minimizes the immediate loss of water to the drains during irrigation, yet allows the rapid removal of perched water after irrigation. The direct loss of irrigation water to the drain during irrigation not only wastes water, increasing the recycling needs, but can also increase the risks of channel collapse, and leg slot and fissure erosion. Rapid water entry into the channel can induce roof collapse and channel erosion, and in situations where the water is of good quality and the soil at moling depth has a sodic tendancy, the risk of channel collapse through slurry failure and dispersion is greatly increased. To achieve the desired flow regime, direct water flow through the leg fissure and slot system must be eliminated or severely throttled. Consideration must, therefore, be given at mole drain installation to the following:

a. minimizing the size of the leg slot and fissures by adjusting leg geometry and selecting appropriate timing to achieve optimum soil moisture conditions
b. sealing off the leg slot and fissures close to the surface through crack disruption and soil recompaction, using techniques such as the trailing wedge system
c. throttling water inflow into the channel at the leg slot/channel interface through the appropriate selection of foot and expander sizes.

Leg slot and fissure sealing at the surface is particularly appropriate in situations where the surface layers are to be left undisturbed, such as in grassland. Its effectiveness in the longer term may, however, decline, if shrinkage cracks develop during longer dry periods to re-open the connection with the unsealed leg slot below. Soil swelling on rewetting tends, however, to offset this effect by closing these connections again. One system, used in the United Kingdom in grassland in the past to prevent the leg slot opening directly to the surface, was to plow out an unbroken furrow slice ahead of the mole plow. The furrow slice was then returned on top of the leg slot after moling, effectively closing the direct connection between the slot and the surface. Such a system, particularly if accompanied also by fissure closure, could in the irrigated situation significantly increase the length of water flow path from the surface to open fissures below, to reduce direct deep percolation losses if shrinkage and cracking occurred.

Throttling the inflow at the leg slot/channel interface will have implications for the rate of drawdown of the perched water table, tending to reduce it. Leg slot sealing at this interface is, however, rarely complete and other fissures in the channel wall are usually present for water entry. In addition, due to the low cost of the operation, very close drain spacings can be economically justified to help achieve the desired discharge rates whenever required. Soil loosening above the channel is a further measure to increase inflow rate, this operation reducing the length of the low permeability flow path into the channel. To avoid damaging the channel, however, any soil loosening operation must be executed after channel formation and should not extend closer than 100 mm to the channel roof.

A further additional or alternate measure to reduce irrigation water loss is to prevent drain outflow during each irrigation by closing the outfall. This method will only be satisfactory in situations where the mole channel is capable of withstanding the added stress of water ponding and where channel grades are relatively low.

In situations where, due to the dispersive nature of the soil, instability may arise in the mole channel even with leg slot and fissure closure, other measures will be required to provide stability. These could include the injection or placement of materials such as gypsum in the vicinity of the channel at regular time intervals to increase the electrolyte concentration of the drainage water and to replace exchangeable sodium, the major cause of dispersion.

C. Reclamation of Saline and Saline Alkali Soils

The leaching of excess soluble salts and the removal of unwanted exchangeable sodium, using standard pipe or open ditch drainage systems, can be an extremely expensive and time consuming process in many fine textured soils of low hydraulic conductivity. Cost rather than leaching need often dictate the drain spacings adopted in these circumstances and the result is a slow non-uniform leaching process. Mole drainage offers a much lower cost alternative and has shown significant promise in smaller scale field trials in Egypt (Cronin, 1982; Spoor and Leeds-Harrison, 1986; Spoor et al., 1990) and in Canada (Sommerfeldt and Chang, 1986). Particular attention needs to be paid to achieving the appropriate water flow regime and to maintaining channel stability in dispersive soil situations.

The water flow regime required for maximum salt removal and uniform leaching with surface water ponding is that shown in Figure 12, where all direct fissure connections between the surface and the drain are sealed off. This is in contrast to the situation with preferential local flow through leg slot and fissures illustrated in Figure 2. Figure 13, taken from Cronin (1982), shows the differences in soluble salt concentration of the drainage water, in mole drain installations installed with minimal and strong leg fissure development, during leaching trials on a saline alkali clay soil. The higher salt concentration in the mole installation with minimal fissure development indicative of greater salt removal is very obvious and Figure 14 shows for the same installation the relative uniformity of leaching between the drains. The mole drains were

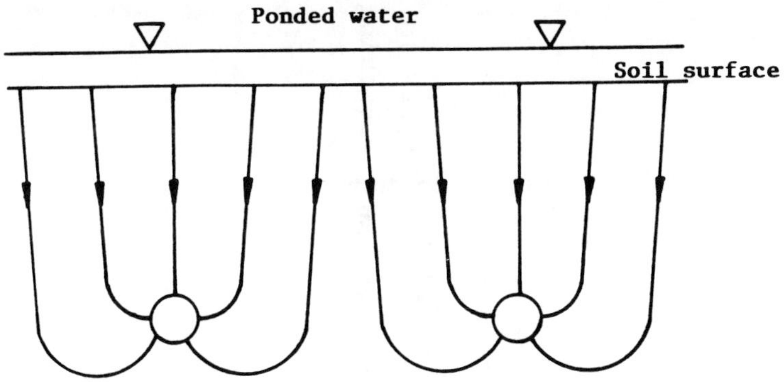

Figure 12. Water flow path required in a reclamation situation.

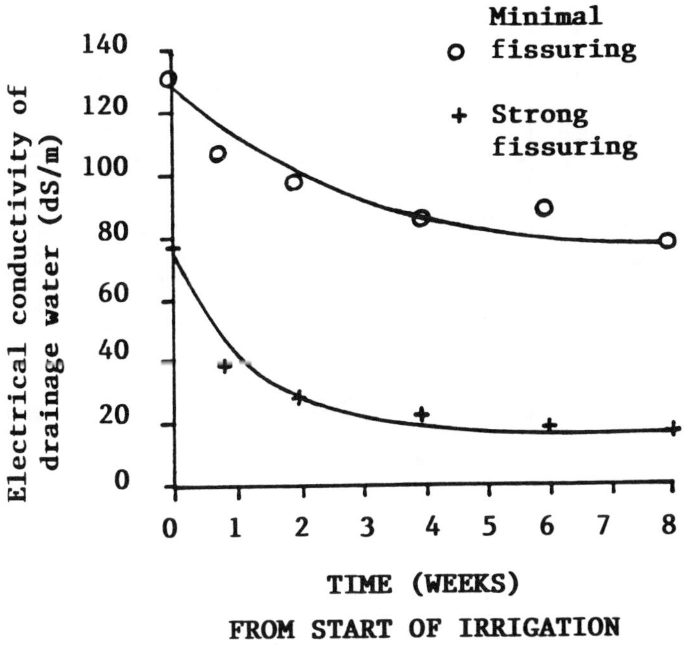

Figure 13. Influence of soil fissure development on changes in salt concentration of drainage water during leaching period.

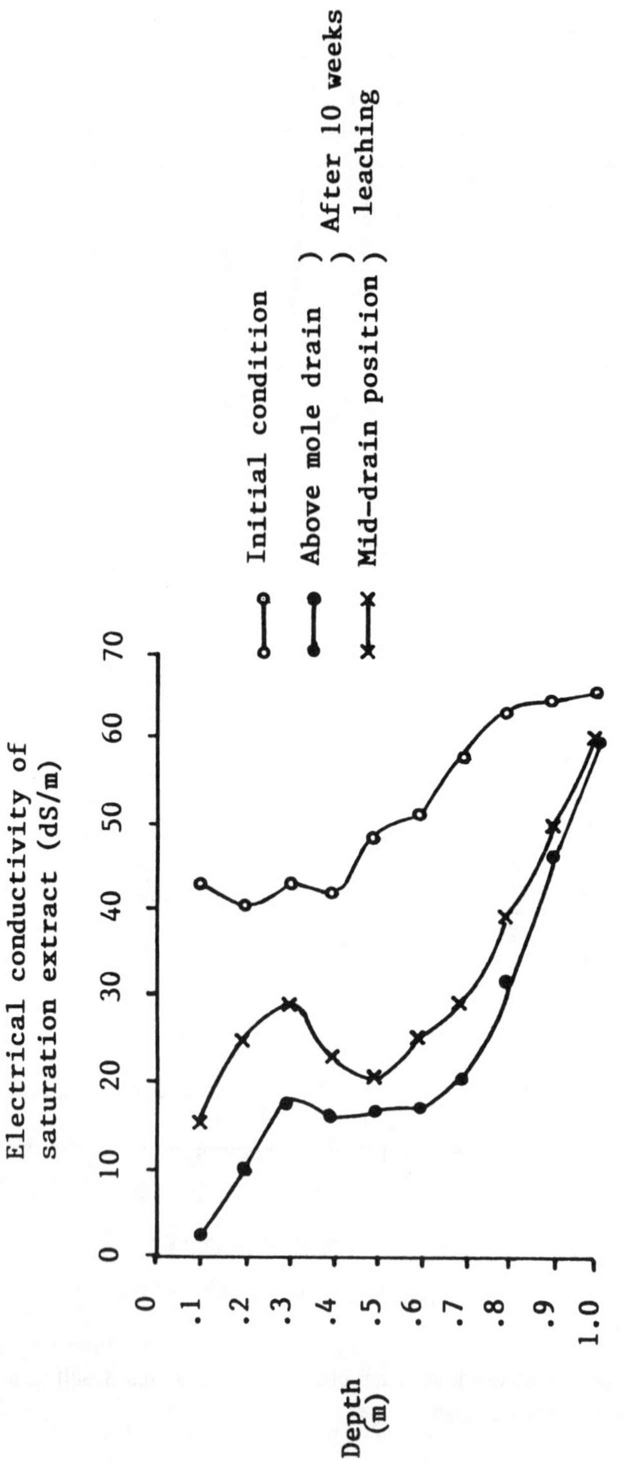

Figure 14. Changes in soil profile salt concentration.

installed at 0.7 m depth and 1.8 m spacing and it can be noted that significant leaching occurred below drain depth.

Channel stability was achieved in the Cronin trials by preventing the direct access of good quality irrigation water into the mole channel. The mole channels were very stable over the 7 month period of the trial, in a highly erodible and dispersive smectitic clay soil containing 90% clay. Initial electrical conductivities of the saturation extract were in excess of 60 dS m^{-1} and the exchangeable sodium percentage ranged between 40 and 65. The direct introduction of irrigation water into the mole channel produced a complete slurry failure within 20 minutes. Channel stability and successful reclamation in this extremely difficult situation indicates the potential of the moling technique for reclamation work. Experience to date in the application of mole drainage for leaching purposes indicates that for success attention must be given to the following procedural and installation factors:

a. Soil pre-treatment
   - (i) establish appropriate moisture conditions for installation
   - (ii) on shrinking soils, install mole drains as soon as the surface is trafficable after wetting, while natural shrinkage crack development is still relatively shallow. This minimizes the risk of connections developing between these cracks and the mole leg fissures at depth with further drying later.

b. Mole channel installation
   - (i) seal off the leg slot and fissures at the surface
   - (ii) close or throttle the connection between the mole channel and the leg slot and fissures above, by selecting appropriate foot and expander sizes
   - (iii) avoid reverse grades in the mole channel to minimize water ponding stresses
   - (iv) if the moles are drawn from open ditches in shrinking soils, close the end of the mole channel at the ditch, to prevent drying and cracking in the channel wall near the outfall during any aging period. This helps avoid local roof collapse.

c. Leaching water application
   - (i) allow an aging period after installation for channel stabilization before wetting up. This aging time must be curtailed, however, if deep shrinkage cracks develop within the soil mass. Wetting must commence before any natural surface initiated cracks link up with the mole channels or lower leg fissures and cracks at depth. Shrinkage crack development can often be slowed down by a shallow surface cultivation during the soil pre-treatment stage, to produce a loosened surface layer.
   - (ii) on soils very susceptible to erosion, leg slot and fissure closure must be complete to avoid piping failures following surface water application. As the mechanical sealing of every slot and fissure connection is not practicable, extreme care is needed at the first

wetting to avoid erosion. The safest and most satisfactory water application system for the first irrigation is controlled overhead sprinkler. The low application rate and unsaturated water movement without surface flow allows soil swelling to seal off any remaining surface fissures. Normal flood irrigation can be reverted to after this initial wetting with minimal risk.

The less structurally stable a soil, the greater the attention needed to these procedural and installation factors. Loosening the upper soil layers, with the addition of chemical amendments where necessary, after mole installation, would assist in speeding up the rate of reclamation in soils with exceptionally low hydraulic conductivities. Care will be needed if this practice is to be adopted to ensure satisfactory closure of the leg slot/mole channel connection.

D. Salinity and Groundwater Control in Irrigated Areas

The normal practice adopted for the long term control of root zone salinity in irrigated areas is regular downward leaching, removing the leaching water continuously through subsurface drains. While this practice is satisfactory in situations where the necessary drainage system can be installed at an affordable cost, cost problems arise in soils of very low hydraulic conductivity, due to the need for very close drain spacings. An alternate procedure in these latter situations, particularly where the rate of salt build up with time is not exceptionally high, would be to adopt an alternating reclamation/salt build-up system. A temporary mole drainage system could be used for reclamation to reduce salinity levels to well below critical limits. This drainage system would then be abandoned until such time as salt levels rose again, when new moles would be installed for reclamation leaching purposes. Such a practice would also be advantageous where the maintenance of a continually functioning mole drainage system is difficult or expensive due to poor soil structural stability.

Mole drains have rarely been used in the past for the control of groundwater tables, a practice which would entail long periods of wetting. Moles in stable structured soils have, however, been shown to be capable of withstanding prolonged waterlogging without collapse (Harris, 1984). Their application for groundwater control in irrigated areas is, nevertheless, likely to be very limited, although there may be some special circumstances where possibilities exist.

The most promising situations for mole drain groundwater table control would be those where soil hydraulic conductivity and specific yield/drainable porosity are low and where soil structural stability in the mole channel is high. The low conductivity and porosity would throttle water inflow rate, hence reducing the risk of significant water depth in the channel. Experience indicates that mole channels are less liable to collapse when water is filled to only one third or less; increased stability problems arise once the channels become more than half full. The soil within the channel must be stable to resist excessive swelling and this is possible in inherently stable structured soils, and in situations with high soil water electrolyte concentrations. The latter is likely to be most common in irrigated areas with saline groundwater. One field situation approaching these required conditions would be where the moles are installed in the confining layer of a semi confined aquifer, with a slow upward flow of saline groundwater. Moles would be unnecessary in unconfined aquifers, since the higher permeabilities would allow the use of wider spaced pipe drainage systems.

Mole drain installation in these situations must be done in the absence of free water and an aging period would be essential to enable the channels to strengthen. The direct ingress of good quality surface water would need to be prevented.

E. Short-Term Dewatering

There are numerous instances where benefits can be achieved by reducing soil moisture contents rapidly for relatively short periods of time. Examples include:

a. dewatering flooded soil surface layers following serious compaction damage, caused by intensive mechanization operations during wet periods
b. drainage to improve the timeliness of operations, allowing earlier cultivations, crop establishment or grazing
c. drainage during one or more cropping periods.

In the surface dewatering case, drainage may only be needed for a few days to enable tillage operations to commence; for improved timeliness, the drainage need may extend from weeks to months. One application in the U.K., is to improve grassland drainage during the winter period, to achieve 1-3 weeks earlier grazing in spring. Drainage for cropping periods may extend from months to one year. In all such cases the installation of a "temporary" drainage system can be fully justified providing it is cost effective.

Mole drainage is adaptable to all these types of situation and can be practiced on soils and under conditions which normally would be considered unsuitable, due to the relatively short life of mole channels in them. Typical applications in "unsuitable" situations would include: moling soils of marginal to "unsatisfactory" texture; moling at shallower depths than usual to match the tractor power available; extending the length of mole runs where closer spaced outfalls are unavailable and using moled main systems on marginal soils. The flow regimes and hence installation techniques required will depend upon the particular application.

F. Wetland Rice

In certain wetland rice situations, benefits can be achieved by rapidly draining down the paddy at specific growth stages such as prior to flowering, and prior to harvest to improve soil bearing capacity at harvest. The water flow regime required is for preferential flow through the leg slot area into the mole channel for rapid surface water removal, with the ability to prevent drainage during flooding periods. In China where such control is practiced in a rice/wheat rotation (Soong and Wei, 1985; Guo et al., 1992), the timing of mole installation and soil structural stability are critical for success.

Soil moisture conditions at moling depth below the puddled layer after the rice harvest and before wheat establishment are effectively saturated and, while suitable for channel formation, they are not conducive for leg fissure development. Conditions after the dryland crop are frequently too dry for stable channel formation, but suitable for leg fissuring. The practice adopted is to mole after rice, forming the channel with a weakly developed leg slot and slight to zero fissuring. Soil shrinkage during the following dryland crop opens up the leg slot and fissures, and the aging period assists in stabilizing the mole channel. A stabilized channel with open leg slot is therefore present at the beginning of the rice cropping season, to provide the required water flow regime.

To prevent water loss during flooding periods, all drain outlets are closed. The success of mole drainage under these extreme stress conditions is almost entirely due to the stability of the soil structure, which is capable of resisting breakdown and excessive swelling even when subjected to prolonged wetting and submergence. The annual drying during the dryland crop re-establishes an open leg slot prior to the next rice season.

The application of mole drainage in other wetland rice situations will be dependent upon the structural stability on wetting of the soils on site.

G. Soil Structure Improvement and the Alleviation of Impeding Layers

The most common approach to alleviating problems of soil compaction, the effects of impeding layers and soil structural damage is to reduce soil density through soil loosening operations. Such action improves drainage, aeration and root development, all of which assist in overcoming the problems. One drawback, however, of these loosening operations, is that the open soil condition left is particularly vulnerable to recompaction by subsequent machinery traffic and grazing animals. In addition, if there is no subsurface outlet for drainage water, the loosened zone behaves like a water sump and the resulting condition can be worse than the initial one. The mole plow offers an alternate approach which may prove more satisfactory under certain conditions.

The vertical fissures and leg slot generated by the mole plow are much more resistant to collapse under vertical loading than the more horizontal cracks developed by loosening equipment. In addition, providing there is an adequate outfall, the mole channel provides a subsurface drain to remove excess water. The leg slot and fissures readily break through compacted and impeding layers improving drainage and aeration. The immediate benefits to root development may in some circumstances be limited to a much more local area than following a loosening operation. This is not usually a major concern in swelling and shrinking soils, since this local improvement rapidly spreads into the undisturbed zone between the mole runs, through shrinkage crack development and root exploration. Similarly in situations where surface or perched water have been the major factors controlling production, the improved drainage provides improvement over the entire soil profile (Jayawardane and Blackwell, 1985). The mole plow approach thus offers important advantages for alleviating impedance and structural problems in situations with adequate drain outfalls and where the soil exhibits some swelling and shrinkage properties.

In this application the mole runs can be spaced much closer together and mole plow foot and leg geometry adjusted to suit working depth and leg fissuring requirements. On sloping sites where mole channel outfalls are poor, moles should be run around the contour to avoid water blow-outs. The risk of wet sumps developing in the fissured zone after moling is minimal, compared with the risks in an undrained loosened area. The moling technique is also particularly appropriate for providing a direct water flow connection between the surface layers, through a restricting layer, to a more permeable layer at depth.

VI. Development of Mole Drainage Systems for New Situations

While the application of mole drainage to problems of short-term dewatering, structure improvement and overcoming impedance will most probably only require the transfer of existing techniques, its successful application in irrigated areas and in currently marginal situations will require technique development. A common weakness in many past unsuccessful mole drainage trials in new situations, has been the failure to first adapt the technique to suit the new conditions before evaluating the results. As a result, it is unclear whether failure was due to a completely impossible situation or whether with modification a working system could have been developed. A systematic approach is outlined below to enable the potential of mole drainage in new situations to be evaluated, and appropriate equipment and techniques to be developed.

A. Systematic Approach to Technique Development

The suggested stages and action required in a developing a moling technique in a new situation are as follows:

a) Preliminary soil tests

Assessments of the following are required:

(i) soil texture and its uniformity or variation with depth
(ii) swelling tendencies
(iii) aggregate stability on wetting
(iv) likely moisture conditions at installation
(v) subsequent wetting and drying pattern and flow requirements.

These assessments will enable decisions to be made on the likely ability of the mole channel to accept good quality water without collapse, the potential flow regime requirements, promising moling depths and potential channel failure mechanisms.

b) Achievement of desired flow regime

Field tests to identify appropriate leg geometries and leg slot and fissure closure techniques where required, to produce the required leg fissuring and water flow regimes.

c) Production of stable mole channels

Through field tests:

(i) select appropriate foot and expander sizes to ensure they are working below their critical depth when at moling depth

(ii) identify channel failure mechanisms, if any, following channel installation with desired water flow regime conditions and after appropriate aging period.

(iii) modify mole plow geometry and/or soil moisture conditions at installation as necessary, to try to counteract the failure mechanisms while still achieving the desired water flow regime

(iv) assess the potential of additional channel stabilizing techniques where necessary

d) Evaluation of mole drainage technique

Once the desired water flow regime and channel stability have been achieved, evaluate the effectiveness of the technique in achieving the desired objectives.

e) Development of practical field system

Develop the technically proven technique, into a practical field system, identifying optimum lengths of run, channel gradients, types of outfall, layouts to fit in with farming system, types of equipment and installation costings to enable benefit/cost analyses to be made.

B. Experimentation, Monitoring and Evaluation Procedures

The experimentation needed to identify and achieve the desired flow regime and stable mole channels can be carried out very successfully on relatively short (20 m length) mole runs. A specialist mole plow although desirable is not necessarily required, but the design of the mole plow legs, feet and expanders used must be in accordance with those used in other successful moling areas, as described in this text. It is also essential to ensure that the mole channels are installed with the mole foot running parallel to the desired grade, that there are no reverse grades or sudden grade changes and that the outfalls are good and suitably protected against erosion or damage.

Monitoring techniques can include direct observation, outflow monitoring and soil water potential and moisture content measurements. Direct observation of the mole installation is probably the single most important investigative technique which should be employed when assessing the performance of the mole channel and leg cracks. This can be undertaken by excavation at right angles to and along one side of the channel, to expose the channel and vertical face as shown in Figure 1. Leg crack size or closure and mole channel condition can then be readily observed. Observation immediately after installation and again after a few days or weeks drying can confirm whether cracks have opened up in the channel and leg slot area, or whether a seal has been achieved.

Longer term and repeated observations can be made without disturbance, by inserting an access tube into the drain and using a borescope to view the channel (Leeds-Harrison et al., 1983). The installation of the access tube is most readily achieved by inserting the tube down behind the plow leg at the time of moling. Access tube diameter needs to be a little less than the thickness of the mole plow leg. Such observations give a very clear picture of the nature of any deterioration in the mole channel.

A rapid assessment of relative differences in channel stability following different treatments can be made by flooding the channel soon after formation. Isolating a 1 m length of channel and flooding with water of appropriate quality induces severe failure stresses, which can assist in quickly discriminating between treatments.

Outflow from individual mole channels can be measured using weirs or tipping bucket flow meters, but, due to the close spacing of the channels, the outflow quantities may be small, hence flow meters capable of dealing with low flows are required. Alternatively the mole channels can discharge collectively into a collector drain which in turn is monitored. This latter technique can be very successful in determining the overall performance of a field system, but occasionally deep seepage water may enter the collector drain making it difficult to separate mole drain flow from deep seepage. The hydrograph from such measurements is a good indicator of drainage scheme performance. With a good connection between the leg fissures and the channel, the discharge response to individual rainstorms should be very rapid. If the fissures, channel connection or channel deteriorate, or are sealed off, longer drain flow recession times and a distinct reduction in the peak flow from the drain may be observed, see Figure 3. Linked with borescope observations, the actual cause of any changed discharge can be attributed to the cracks or channel or both.

To obtain information concerning the complete water flow system, measurement of soil water potential can be made using arrays of tensiometers. Particular care must be taken when installing tensiometers in this situation to ensure the ceramic tip of the tensiometer monitors the desired water flow, i.e., water flowing in the macropores and/or within the clay matrix. Installation for assessing macropore flow should proceed by placing the tensiometer in an augered hole, ensuring the tip of the device is in contact with the matrix. Dry sand may then be placed around the remaining exposed ceramic and the augered hole backfilled with bentonite. The dry sand around the ceramic should be left unwetted for a week during which soil water in the vicinity of the tensiometer tip may equilibrate with the sand. This usually results in small cracks opening up between the natural structure of the soil and the sand. The tensiometer can then be filled with water in the usual way. Rapid soil water potential fluctuations during periods of drain flow are indicative of a direct connection between the tensiometers and macropores and vice versa.

Dye tracing techniques are useful to study the preferential pathways for water movement to mole drains. Choice of dye is important as some dyes are very strongly absorbed on to the clay matrix and hence are unsatisfactory. Methylene Blue, Rhodamine B, Lissamine FF and Fluorescene have all been used by various workers but none of these can as yet be considered as a perfect dye for this purpose (Smettem and Trudgill, 1983; Hallard and Armstrong, 1992). A very weak gypsum paste has also been used by some workers at a dilution of 5:1 water to gypsum and has proved useful in identifying the large conductive pores but not the finer macropores (Fitzpatrick et al., 1985).

Numerous tests have been devised to assess aggregate stability in mole drainage situations. The most common methods are the basic Childs (Childs, 1942a) and modified Childs test (Rycroft and Alcock, 1974; McGregor and Stevens, 1977), where stability is assessed in terms of the differences in the drainage moisture characteristic curves following slow and rapid wetting of 1 to 2 mm diameter air dry aggregates; wet sieving tests such as Yoder (1936); aggregate stability test of Emerson (1967) and a dispersion and slaking test developed by McQueen (1981), further modified by Hann, to assess the stability of remolded soil (M.J. Hann, 1993, personal communication). These latter tests have the added advantage of being rapid to execute.

VII. Conclusions

Mole drainage is currently restricted to a very limited number of soil, climatic and drainage situations. Attempts to broaden its application in the past, without detailed attention to the equipment and installation techniques used, have generally not been particularly successful.

Any successful extension of mole drainage will firstly require the adaptation and development of the technique to suit the new requirements. This will frequently involve modifications to the equipment and installation technique, to achieve the desired water flow regime and stable mole channels. A systematic approach is available to assist in this development, to bring the technique to the stage where a fair evaluation can be made of its potential, for helping overcome a particular subsoil problem.

While mole drainage will not be adaptable to all heavy soil situations, with careful development the technique has potential for extension well beyond current limits.

References

Blackwell, J., N.S. Jayawardene, and R.K. Butler. 1990. Design, construction and preliminary testing of a slotting implement for gypsum enrichment of soil. *J. Agric. Engng. Res.* 46:81-92.

Bowler, D.G. 1980. *The drainage of wet soils.* 1st. ed. Hodder and Stoughton.

Castle, D.A., G. Spoor, E.O. Onasanya, and A.K. Ormandy. 1992. Mole plough and soil factors influencing the hydraulic connection between mole channels and the permeable backfill placed over pipe drains. *J. Agric. Engng. Res.* 51:217-235.

Childs, E.C. 1942a. Stability of clay soils. *Soil Sci.* 53:79-92.

Childs, E.C. 1942b. The mechanics of mole drainage. *Empire J. Expt. Agric.* 10:169-181.

Cronin, C.J. 1982. Mole drainage for the reclamation of fine textured saline-alkali soils. M.Sc. Thesis, Silsoe College, Cranfield Institute of Technology.

Dennis, C.W. and S.G. Croote. 1977. The minimum size of permeable fill used with mole drainage. *Agric. Water Management* 1:143-153.

Emerson, W.W. 1967. A classification of soil aggregates based on their coherence in water. *Aust. J. Soil Res.* 5:47-58.

Eriksson, J., G Berglund, and I. Olovsson. 1987. Drainability and drainage of heavy clay soils: studies of mole drainage in Sweden. p. C7-C13. Proc. 3rd International Workshop on Land Drainage, Ohio State Univ., Columbus, Ohio. 7-11 Dec.

Fitzpatrick, E.A., L.A. Mackie, and C.E. Mullins. 1985. The use of Plaster of Paris in the study of soil structure. *Soil Use and Management* 1:70-72.

Fouss, J.L. 1965. Plastic drains and their installation p. 55-57. In: *Drainage for efficient crop production.* Conf. (Chicago). Amer. Soc. Agric. Eng., St Joseph, Michigan.

Gebre-Michael, G.T. 1989. Closure of mole plough leg slot and associated fissures for reclamation purposes. M.Sc. Thesis, Silsoe College, Cranfield Institute of Technology.

Godwin, R.J., G. Spoor, and P.B. Leeds-Harrison. 1981. An experimental investigation into the force mechanics and resulting soil disturbance of mole ploughs. *J. Agric. Engng. Res.* 26:477-479.

Guo Guoshuang, Yugi Hu, Baozhong Pau, and Zugui Liun. 1992. Research on irrigation-drainage technique for water-logged paddy field. Inter. Workshop on Soil and Water Engineering for Paddy Field Management. Bangkok.

Hallard, M. and A.C. Armstrong. 1992. Observations of water movement to and within mole drainage channels. *J. Agric. Engng. Res.* 52:309-315.

Harris, G.L. 1984. Effect of mole submergence on the life of mole channels. *Agric. Water Management* 8:361-374.

Hopewell, H.G. 1953. Mole drainage investigations in New Zealand. 111. Investigation to determine the effect of increasing the thickness of the mole plough blade or knife. *New Zealand J. Sci. Tech.* 35A:237-249.

Hudson, A.W., H.G. Hopewell, D.G. Bowler, and M.W. Cross. 1962. The draining of farm lands. 2nd ed. Bulletin No 18. Massey College, New Zealand.

Jayawardane, N.S. and J. Blackwell. 1985. The effects of gypsum-enriched slots on moisture movement and aeration in an irrigated swelling clay. *Aust. J. Soil Res.* 23:481-492.

Leeds-Harrison, P.B., G. Spoor, and R.J. Godwin. 1982. Water flow to mole drains. *J. Agric. Eng. Res.* 27:81-91.

Leeds-Harrison, P.B., R.K. Fry, C.J. Cronin, and J.E. Gregory. 1983. A technique for the non-destructive monitoring of subsurface drains. *J. Agric. Eng. Res.* 28:479-484.

Leeds-Harrison, P.B., C.S.P. Shipway, N.J. Jarvis, and E.G. Youngs. 1986. The influence of soil macro-porosity on water retention, transmission and drainage in a clay soil. *Soil Use and Management* 2:47-50.

McGregor, A.J. and J. Stevens. 1977. Research and Development Report, Project No 15/2, West of Scotland Agric. College, Auchincruive, Ayr, Scotland.

McQueen, D. 1981. A new method of aggregate stability measurement. *New Zealand Soil News.* 29:127-132.

Muirhead, W.A., E. Humphreys, N.S. Jayawardane, and J. Moll. 1993. Effect of subsurface drains on soil water conditions in an irrigated vertisol with a high water table. (In press.)

Mulqueen, J. 1971. Mole drain failure due to drawing moles downhill. *Irish J. Agric. Res.* 10:241-244.

Mulqueen, J. 1985. The development of gravel mole drainage. *J. Agric. Engng. Res.* 32:143-151.

Nicholson, H.H. 1942. The principles of field drainage. 1st ed. Cambridge Univ. Press.

Ormandy, A.K., 1985. Improvements to mole plough design. M.Phil. Thesis, Silsoe College, Cranfield Institute of Technology.

Rycroft, D.W. and M. Alcock. 1974. The assessment of suitability of soils for moling. Tech. Bulletin No 74/1. Field Drainage Expt. Unit, MAFF, Cambridge, UK.

Schwab, G.O. 1955. Plastic tubing for subsurface drainage. *Agric. Eng.* 36:86-90,92.

Schwab, G.O., N.R. Fausey, and D.W. Mitchener. 1974. Comparison of drainage methods in a heavy-textured soil. *Trans. ASAE* 17:424, 425, 428.

Smettem, K.R.J. and S.T. Trudgill. 1983. An evaluation of some fluorescent and non-fluorescent dyes in the identification of water transmission routes in soils. *J. Soil Sci.* 34:499-509.

Sommerfeldt, J.G. 1984. Performance of unlined and lined mole drains in a saline clay loam field. *Can. Agric. Eng.* 26:1-5.

Sommerfeldt, J.G. and C. Chang 1986. Desalinization of an irrigated, mole-drained, saline clay loam soil. *Can. J. Soil Sci.* 67:263-269.

Soong, Si-Tu and Zhang Wei. 1985. Subsurface drainage of lowland rice fields in China. p.351-366. In: *Soil Physics and Rice.* IRRI, Los Banos, Philippines.

Spoor, G. 1987. Potential ways of increasing mole channel stability. p. C79-C83. Proc. 3rd International Land Drainage Workshop. Columbus, Ohio. 7-11 December.

Spoor, G. and R.J. Godwin. 1978. An experimental investigation into the deep loosening of soil by rigid tines. *J. Agric. Engng. Res.* 23:243-258.

Spoor, G., P.B. Leeds-Harrison, and R.J. Godwin. 1982. Some fundamental aspects of the formation, stability and failure of mole drainage channels. *J. Soil Sci.* 33:411-425.

Spoor, G. and P.B. Leeds-Harrison. 1986. Extension of mole drainage practices. p 443-453. In: K.V.H. Smith and D.W. Rycroft (eds.) Proc. 2nd. International Conference, Hydraulic design in water resource engineering: Land Drainage, Southampton. Publ. Springer-Verlag.

Spoor, G. and R.A. Ford. 1987. Mechanics of mole drainage channel deterioration. *J. Soil Sci.* 38:369-382.

Spoor, G., R.J. Godwin, and S.M. Miller. 1987. Mole plough grade control. *J. Agric. Engng. Res.* 38:145-166.

Spoor, G., M.J. Hann, and A.S. Centeno. 1989. Influence of mole plough leg and expander geometry on soil disturbance. p. 1639-1646. In: V.A. Dodd and P.M. Grace (eds.) Agric. Eng. Proc. 11th Congress on Agric. Engng., CIGR, Dublin. 4-8 September. Publ. A. Balkema.

Spoor, G., C.J. Cronin, and P.B. Leeds-Harrison. 1990. Mole drain installation for leaching purposes. Section 3, p. 47-54. In: Proc. of Symposium on Land Drainage for Salinity Control in Arid and Semi Arid Areas. Cairo.

Spoor, G., R. Earl, and P.N. Wheeler. 1992. Development of lower cost drainage systems for heavy soils. Home-Grown Cereals Authority, Report No 64. London.

Weil, C., S. Natho-Jina, R. Chambers, and K. Wires. 1991. Mole drainage in silicate clay soils subject to freezing. *Trans. ASAE* 34:1693-1698

Yoder, A. 1936. A direct method of aggregate analysis of soils and a study of the physical nature of erosion loss. *J. Amer. Soc. Agron.* 28:337-351.

Slotting – A Deep Tillage Technique for Ameliorating Sodic, Acid and other Degraded Subsoils and for Land Treatment of Waste

N.S. Jayawardane, J. Blackwell, G. Kirchhof
and W.A. Muirhead

1-56670-020-5/95/$0.00+.50

I. Introduction

The need to provide food and clothing for the increasing world human population has necessitated extension of agriculture to soils of low productivity with adverse subsoil properties, especially in developing countries. These problem soils include sodic, acid and compacted soils. Excessive trafficking with heavy farm equipment, indiscriminate fertilizer use and other poor irrigation and agronomic management practices have also contributed to the development of these subsoil problems on existing agricultural lands, especially in developed countries.

Conventional tillage techniques such as deep ripping have generally not provided long term improvements of these problem soils. Thus heavy clay soils and compacted soils can be ameliorated by plowing or ripping to a uniform depth, sometimes combined with addition of gypsum to correct sodicity. However, the beneficial effects of the increased macro-porosity created by these conventional tillage practices are often lost within a few cropping seasons due to recompaction under trafficking. Slumping through leaching of the applied gypsum also leads to loss of macro-porosity (Loveday et al., 1970). In dryland cropping, loosening of compacted soils by tillage can lead to short-term increases in yields of annual crops, but it can also accelerate soil loss through erosion leading to long-term yield reductions (Bentley, 1985). In acid soils, incorporation of lime using conventional tillage practices is an effective method for correcting surface acidity, but it is ineffective in correcting subsoil acidity, because of poor mixing of the lime with the soil to depth (Adams, 1981). Subsoil acidity cannot be corrected by surface lime application due to the low solubility and mobility of lime. In addition, large quantities of lime or gypsum are required to correct highly acidic and sodic subsoils uniformly to depth.

A more promising approach to obtaining long-term improvements of these problem soils with adverse subsoil conditions is to use new tillage techniques, which cause disturbance and addition of amendments to a part or segments of the soil, leaving the rest of the soil profile undisturbed. These techniques include farming with traffic lanes, bed farming and slotting. The aim of this paper is to review the research on the slotting technique and its potential applications in ameliorating degraded soils. The use of slotting technique for land treatment of rural and urban waste is also discussed.

II. The Slotting Technique

Slotting involves the creation of narrow, parallel bands of tilled soil, loosened to the required depth, leaving the rest of the soil undisturbed (Figure 1). The loosened soil may be mixed with an ameliorant such as gypsum, lime, crop residue, fertilizer or a combination of ameliorants. The ameliorant is chosen according to the soil problem which needs to be remedied (Table 1). Slots may

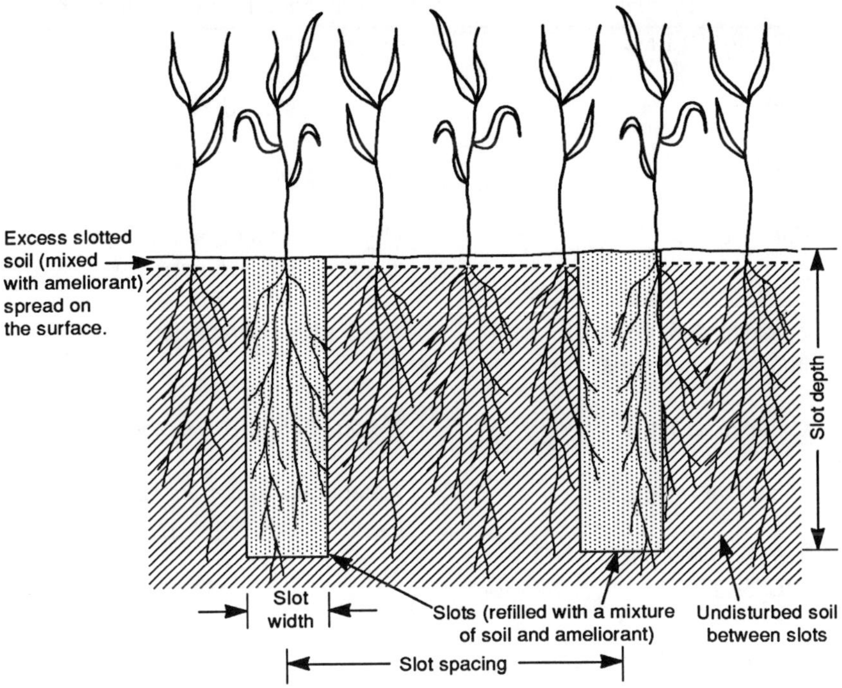

Figure 1. Sketch of soil profile showing slot configuration.

also be used to dispose of urban waste such as sewage sludge on agricultural lands, while at the same time increasing their productivity (Section VIII).

The slotting technique offers flexibility in the choice of the slot configuration, namely, slot width, depth and spacing (Figure 1) which could be used to address specific problems on farms. The optimum configuration depends on the type and extent of the soil problem, as well as the choice of crops and agronomic practices, and the climate at the site (Section XII). In addition, the amount of ameliorant added and its placement in the slot or in combination with surface application can be adjusted to provide optimum solutions to specific soil problems.

Although the soil loosening is restricted to the slot, the improvements in root zone conditions for crop growth extend into the undisturbed soil. This is illustrated with experimental data in sections III to VII on the use of slotting to overcome subsoil problems. The process of slot formation by rotary slotting is described in Section X, and slotting implements are described in Section XI.

Table 1. Choice of Ameliorants for Slotting

Problem	Ameliorant used in Slotting
A. SOIL PROBLEM	
1. Sodic soil	Gypsum, sludge
2. Acid soil	Lime, gypsum, sludge
3. Sandy or heavy clay	Organic residues, sludge, ± nutrients/micro-nutrients
4. Nutrient deficiency of immobile nutrients	Nutrients/micro-nutrients
5. Compacted layer	No ameliorant (or as above)
6. Mine site revegetation	Ameliorant to correct specific deficiencies/toxicities
B. WASTE DISPOSAL AND UTILIZATION	
1. Sewage disposal	
2. Other urban and rural waste disposal	

III. Slotting or Segmental Trenching without Ameliorants to Improve Soils with Hardpans or Claypans

Eck and Unger (1985) reviewed the problem of restricted water movement and root growth through soils with plowpans, fragipans, duripans and claypans. They also examined the effectiveness of profile modification practices such as plowing, ripping, chiseling and trenching to ameliorate these soils under varying climatic regimes and found wide variations in crop responses. They emphasised that the successful application of these tillage techniques depends on careful identification and assessment of the soil factors that limit crop growth on the non-ameliorated soils under the climatic conditions that exist at the site (Letey, 1985) and the extent to which the amelioration techniques overcome these soil limitations. This paper only reviews the research on segmental trenching and slotting.

The effects of segmental trenching or slotting without adding a soil amendment was evaluated on a Harlingen clay soil of low permeability (Heilmann and Gonzalez, 1973). They showed that trenching, (0.13 m wide, 0.6-1.0 m deep, beneath beds spaced 1.0 m apart) reduced soil bulk density in the trench and increased water infiltration, water extraction and root growth. Cotton yields were significantly increased over a 2 year period by trenching.

Deep trenching (0.51 m wide, 1.5 m deep, 1.52 m spacing) passing through a fragipan, without addition of chemical amendments, increased water storage, water extraction, root growth and yield of a sorghum crop (Bradford and Blanchar, 1977, 1980). Yields were further increased by mixing lime, fertilizer and sawdust in the trench. Fritton and Olson (1972) found that refilling the trench with topsoil slowed the reformation of the fragipan. Yushkevich et al. (1982) found that slotting a chernozem soil to 0.35-0.50 m depth at 1.0 m spacing increased water storage, wheat yield and the net profit to farming.

Eck and Taylor (1969) observed an increased rate of water infiltration, depth of wetting and crop water uptake within a wide trench formed by a ditching machine on a slowly permeable clay loam soil. This led to marked increases in yield of grain sorghum, when irrigation was restricted to pre-plant applications. However, with adequate irrigation, grain yields were unaffected. This illustrates the importance of agronomic management and climate in modifying the crop response to tillage practices (Vepraskas, 1988). Polupan (1982) found that trench plowing, which was widely used in the dry steppe zone of the Ukraine, increased porosity and water availability in solonetzic soils. In reviewing the research on soil slotting in the U.S.S.R., Sidorenko et al. (1980) found that in a majority of studies slotting showed beneficial effects in increasing crop yield and soil water retention, and in reducing runoff. They indicated the need for further research to understand the soil-water-plant interactions which lead to increased yields in slotted soils.

IV. Gypsum-Slotting to Improve Sodic Clay Soils

A. Problems of Sodic Clay Soils and their Long-Term Amelioration

Heavy clay soils cover large areas of land used for agriculture around the world. In fine textured soils of the Riverine Plains of south-eastern Australia, which often have sodic clay upper B horizons, productivity of irrigated crops other than rice is often low. Loveday et al. (1970) showed that this was due to low infiltration rates which reduced the amount of water stored and hence the water available for crops during prolonged drying cycles. McIntyre et al. (1982) showed that the low saturated hydraulic conductivity of the upper B horizon reduced the wetting of the lower soil layers (to water contents less than field capacity), even with prolonged ponding. The slow internal drainage of the upper B horizon after irrigation or prolonged rainfall periods also restricts the oxygen supply to the lower layers (Meyer et al., 1985; Jayawardane and Meyer, 1985) leading to reduced crop growth. These soils can be ameliorated by plowing or ripping to a uniform depth over the whole area, combined with gypsum addition. However, the beneficial effects are lost within a few seasons. This is due to recompaction under trafficking and slumping during irrigation.

Jayawardane and Blackwell (1985) developed the gypsum-slotting technique to provide long-term amelioration of these sodic clay soils. The gypsum-enriched slots, with high macro-porosity were expected to provide preferential pathways for rapid water flow and root growth to the lower soil depths, which would increase the water storage and availability. The slots could also increase the internal drainage from the upper B horizon, thus improving soil aeration. The technique was also seen as providing a long-term benefit, as the loosened soil in the slots would be protected from recompaction during trafficking, because the stronger, uncultivated soil adjacent to the slot would support much of the wheel load (Section IX). This contrasts with the complete support of the wheel by the loose soil after extensive deep loosening by deep plowing or ripping, which leads to recompaction. The application of gypsum only to the narrow tilled areas, could increase its concentration and hence its effectiveness in stabilizing the macropore structure of the slots. The results of field tests on the gypsum-plotting technique are described in the following sections.

B. Effects on Water Content Profiles during a Cropping Season

The gypsum-slotting technique was tested on a sodic clay (Mundiwa clay loam) soil, using a sorghum crop grown under irrigation (Jayawardane and Blackwell, 1985). The slots were 0.4 m deep, 0.6 m apart and the gypsum application rate in the slots was equivalent to 8 t ha^{-1} broadcast. The gypsum-slotted soil was compared with a non-ameliorated control. The effectiveness of the gypsum-slotting technique in improving root zone conditions is closely related to the two-dimensional (within and between slot) changes in water content profiles during infiltration, redistribution and crop water extraction. Soil water content (θ_v) change throughout the season were measured using a neutron probe (Jayawardane et al., 1984). In the gypsum-slotted plots, θ_v was measured both in the slots and midway between two slots.

In the non-ameliorated soil, θ_v remained high throughout the season except in the 0.15 m layer (Figure 2). During irrigation, the soil was wetted to a depth of 0.45 m. Water depletion through evapotranspiration and internal drainage occurred only in the 0.15 m layer between sowing and 70 DAS (days after sowing), but extended to a depth of 0.45 m during the prolonged drying cycles later in the cropping season.

In the gypsum-slotted soil, the water infiltration, redistribution and extraction patterns were markedly different (Figure 2). In and below the slots, more pronounced θ_v increases occurred during irrigation, and extended to the greater depth of 0.95 m. This was due to the higher hydraulic conductivity of the slots compared to the adjoining soil resulting in preferential water flow into and through the slots. Between irrigations, very marked decreases in θ_v occurred due to relatively faster rates of water redistribution. Preferential root growth and function in the slots due to lower mechanical resistance (Jayawardane and Blackwell 1990) and better aeration (Jayawardane et al. 1987) which led to

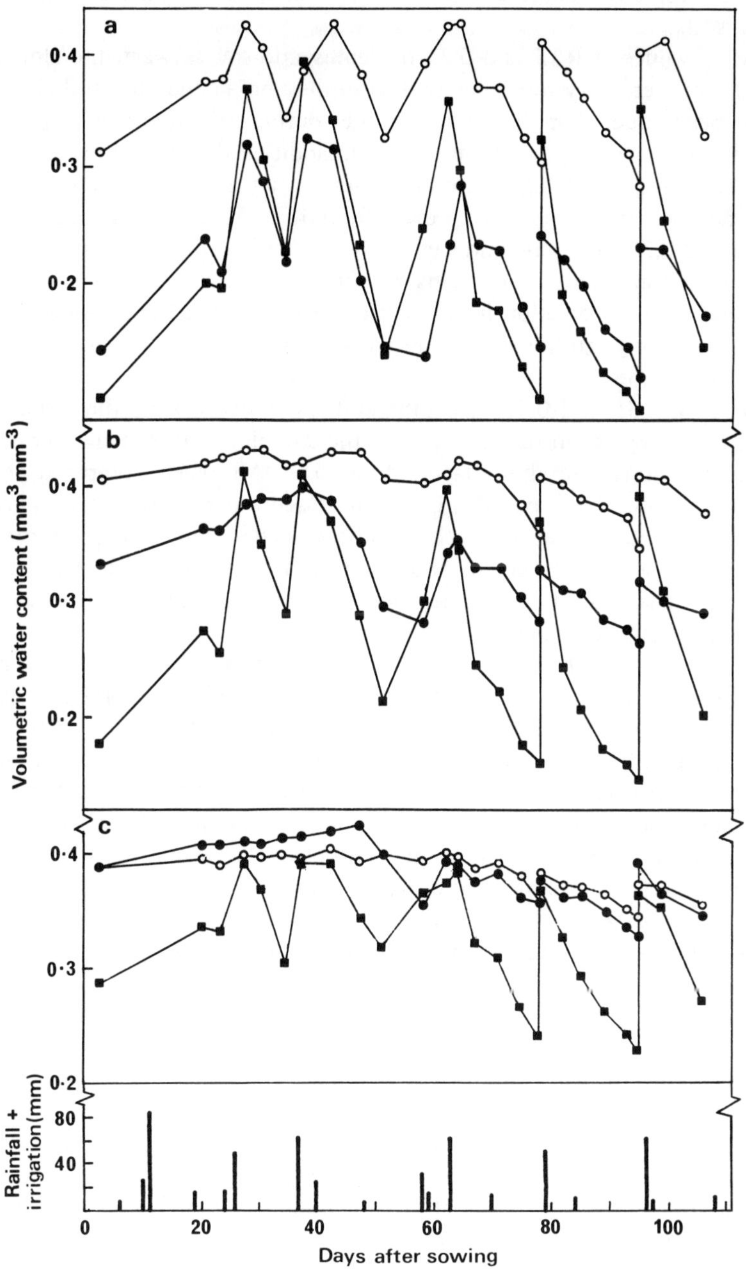

Figure 2. Changes in volumetric water content (θ_v) during the cropping season at (a) 0.15m, (b) 0.25m and (c) 0.45m in the non-ameliorated soil (○) and gypsum-slotted soil within (■) and between (●) slots. (From Jayawardane and Blackwell, 1985.)

increased crop cover could also have contributed to faster water extraction in the slotted soil.

The θ_v values at 0.15 and 0.25 m depths midway between the slots in the slotted plot were very much lower than in the non-ameliorated soil (Figure 2). This is partly due to smaller increases in θ_v during irrigation resulting from the preferential flow of water into the slots. In addition there was greater depletion of water during the drying cycle in the surface layers of the slotted plot. Thus the θ_v decrease during the drying cycle from 38 to 52 DAS is more pronounced at 0.15 and 0.25 m depths midway between the slots than at the corresponding depths in non-ameliorated soil. This may be attributed to greater water uptake by roots due to better aeration in this layer and to increased drainage of water to deeper layers through the slots, from the surface layers (0.25 m) of undisturbed soil between slots.

The θ_v at 0.35 and 0.45 m depths midway between slots did not decrease during drying cycles in the early part of the cropping season. However, marked decreases occurred in these layers from 52 DAS when compared to non-ameliorated soil. This is possibly due to increased water uptake by roots growing more rapidly into this layer, which would have better oxygen supply through the surface layers with relatively lower θ_v, compared to non-ameliorated soil. Similar patterns in changes of θ_v profiles due to gypsum-slotting were observed in subsequent field experiments.

The soil physical factors which directly influence root growth and function and hence crop growth are water availability, soil aeration, soil strength and soil temperature (Cockroft, 1984; Letey, 1985; Miller, 1986). Hence in evaluating effects of any tillage practice on crop growth, the changes in these soil physical factors due to tillage should be monitored at all depths within the root zone throughout the cropping season, especially in the periods when the yield components are affected. Each of these soil physical factors is closely related to θ_v of the soil. Therefore by monitoring the changes in θ_v profiles due to gypsum-slotting, the changes in these soil factors which directly affect crop growth could be assessed, as illustrated for winter and summer cropping in the following sections.

C. Effects on Soil Aeration during Winter Cropping

The effect of the changes in soil water content and hence air-filled-porosity (ε_a) profiles due to gypsum-slotting on growth and yields of a winter wheat crop was studied on the same experimental plots. Through good irrigation scheduling of the wheat crop, root zone limitations due to water unavailability or high soil strength were avoided. Previous studies on these soils have shown that an ε_a value of about 0.08 mm^3 mm^{-3} is required to provide an adequate flow of oxygen to the root zone. The neutron probe counts taken in the non-ameliorated and gypsum-slotted plots were used to measure θ_v, from which the ε_a profile changes throughout the season at each depth was calculated (Jayawardane and Greacen,

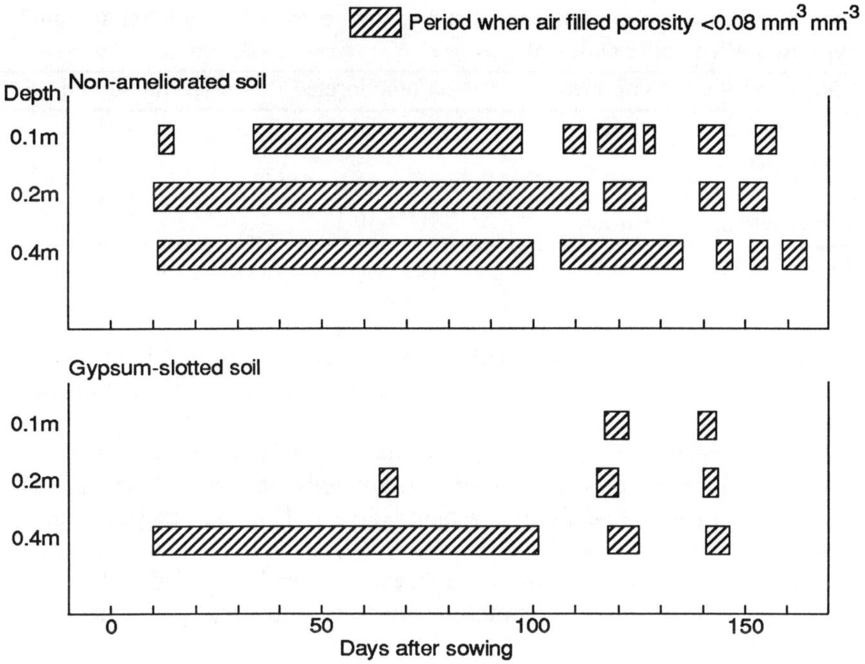

Figure 3. The time period when air-filled-porosity was less than 0.08 mm^3 mm^{-3} at different soil depths in non-ameliorated (control) and gypsum-slotted soil (midway between slots). Within the slot, the air-filled-porosity was always above 0.08 mm^3 mm^{-3}. (From Jayawardane et al., 1991.)

1987). From this data, the period when ε_a is less than 0.08 mm^3 mm^{-3}, leading to restricted oxygen supply, was determined for the two soil treatments (Figure 3).

In the first half of the season, a prolonged period of winter rainfall combined with low evapotranspiration rates caused excessive wetting of the surface layers of the non-ameliorated soil (Jayawardane et al., 1987). This resulted in ε_a values falling below 0.08 mm^3 mm^{-3} during 33-74 DAS at 0.1 m, 33-103 DAS at 0.2 m and 33-120 DAS at 0.4 m depths, respectively. This would cause restrictions in oxygen flow to the root zone. In contrast, in the gypsum-slotted plots the ε_a values were higher than 0.08 mm^3 mm^{-3} at all depths in the slot and at 0.1 and 0.2 m depths midway between the slots. Thus, the improved internal drainage of the gypsum-slotted soil led to adequate aeration of the root zone.

Crop emergence was uniform in both treatments with around 300 plants per m^2 being established. Poor aeration in the non-ameliorated soil during the main tillering period led to the formation of only 0.2 tillers per plant, compared to 2.5 tillers per plant in the gypsum-slotted soil. Tillering increased from 0.2 to 1 tiller

Table 2. Wheat grain yields and yield components in non-ameliorated and gypsum-slotted sodic clay soil

Yields and yield components	Non-ameliorated	Gypsum-slotted
Yield (g/m^2) (12% moist)	541	791
Spikes per m^2	367	513
Kernels per spike	33.9	35.1
Kernel wt (mg) (oven dry)	39.1	39.1

(From Jayawardane et al., 1987.)

per plant in the non-ameliorated soil after soil aeration conditions improved later in the season. However reduced tillering led to reductions in spikes per m^2 which resulted in a yield decrease (Table 2).

The soil aeration conditions were adequate in both treatments during and after flowering, which could explain the similarity in the other two yield components, kernal number per spike and kernal weight (Table 2). However, excessive rainfall during this period could also have caused reductions in these yield components, leading to further reductions in yields on the non-ameliorated soils. In seasons with evenly distributed winter rainfall, smaller crop responses to soil amelioration techniques could be expected.

Comparison of wheat yields along a transect between slots up to 1 m apart showed no relationship between yield and distance from the slot (Jayawardane et al., 1987). However, with further increase in the spacing between slots the water availability and soil aeration conditions midway between the slots will deteriorate, leading to crop yield reductions.

D. Effects on Water Infiltration, Storage and Extraction Patterns during Summer Cropping

On the same sodic clay soil, a new experiment was established in 1986 to evaluate the long-term changes in soil physical properties and summer crop responses to gypsum-slotting and gypsum-applied with conventional deep tillage practices. Macro-porosity and intrinsic permeability at -10 kPa water potential were measured on small cores, after a wheat crop. Deep ripping and slotting with increasing amounts of applied gypsum led to significant increases in macro-porosity and intrinsic permeability (Blackwell et al., 1991). Slots without gypsum had similar macro-porosity and intrinsic permeability to the non-ameliorated soil, due to slaking and slumping within the slot. These two treatments also showed significantly higher run-off during winter rains compared to the deep ripped and slotted plots with gypsum additions (Blackwell et al., 1991). These results indicate the need to chemically stabilise the macropore structure created by slotting or other tillage techniques in sodic clay soils.

Table 3. Soil water depletion (mm) due to evapotranspiration during a drying cycle and the water intake (mm) during subsequent irrigations of a sunflower crop, in non-ameliorated and gypsum-slotted sodic clay soil

Treatment	Water depletions/additions	Irrigation number					
		1	2	3	4	5	Mean
Non-ameliorated	Depletion	18	42	61	87	65	55
	Addition	19	49	52	79	63	52
Gypsum-slotted	Depletion	47	72	84	120	73	79
	Addition	45	72	81	114	62	75

(From Jayawardane et al., 1987.)

The effect of gypsum-slotting on water intake during five flood irrigations, and on subsequent soil water depletion during drying cycles, was examined under a summer sunflower crop in 1987/88 (Jayawardane et al., 1990). Each irrigation event was carried out on the same date on both plots. At each irrigation, the soil water depletion and addition in the gypsum-slotted plots was greater than in the non-ameliorated soil (Table 3). The similarity between soil water depletion during a drying cycle and the amount of water stored during the subsequent irrigation within each treatment is also shown in Table 3. Previous studies have shown that infiltration into swelling clay soils occurs rapidly through crack filling and sorption (Collis-George, 1980). The size, distribution and depth of crack formation is closely related to the extent of soil water extraction during the previous drying cycle, especially in the surface soil layers. In the gypsum-slotted soil, the development of cracks is likely to be enhanced by the increased drying of both the surface soil layers and the slots (Figure 2) due to increased water use by faster growth and development of the crop (Jayawardane et al., 1987). This in turn leads to increased infiltration and reduced runoff (Blackwell et al., 1991). While long term steady infiltration rates are also increased by gypsum-slotting (Jayawardane and Blackwell, 1986; Cook et al., 1989), its relative contribution to infiltration during the short contact opportunity time of summer flood irrigations is small.

The effect of gypsum-slotting on water availability for a maize crop during drying cycles between irrigations was measured during the 1988/89 summer season. The non-ameliorated and gypsum-slotted soils were irrigated six times during the summer, with each irrigation being applied on the same date. Leaf water deficit at mid-day was measured in each treatment just prior to irrigation by taking leaf samples and measuring leaf water content before and after floating in water (Jayawardane et al., 1991). At each irrigation the gypsum-slotted plots had a lower leaf water deficit than the non-ameliorated plots (Table 4), with mean values of 11.9% and 10.1% respectively. The maize yields decreased from 11.5 t ha^{-1} in the gypsum-slotted plots to 7.0 t ha^{-1} in the non-ameliorated plots.

Table 4. Leaf water deficit (% of turgid weight) of a maize crop prior to each irrigation in non-ameliorated and gypsum-slotted sodic clay soil

Treatment	Irrigation number						
	1	2	3	4	5	6	Mean
Non-ameliorated	9.6	10.8	10.2	12.3	13.7	14.9	11.9
Gypsum-slotted	9.3	9.8	8.7	10.0	9.8	12.9	10.1

(From Jayawardane et al., 1991.)

In the next year the yields were 7.5 and 5.4 t ha^{-1}, respectively. Gypsum-slotting also led to increases in wheat yield, in farm trials (Table 5).

On a dense clay soil exposed by landforming Blaikie et al. (1988) found that the photosynthetic efficiency of pasture canopies declined rapidly after 40 mm depletion of soil moisture, but remained high in the gypsum-slotted plot even after 72 mm depletion. The slots occupied 12.5% of the ground area sampled by the gas chamber used for measuring photosynthetic efficiency. They also showed that gypsum-slotting increased rates of evapotranspiration and dry matter production, mainly due to improved growth on or near the slots. At the same site Kelly et al. (1987) showed that pasture dry matter increased significantly from 3.00 t ha^{-1} in non-ameliorated soil to 4.95 t ha^{-1} in gypsum-slotted soil with irrigation at 70 mm depletion. The yield differences were not significant when irrigated at 35 mm depletion, which is consistent with the decline in relative photosynthetic efficiency occurring only after 40 mm soil water depletion in the non-ameliorated soil.

Thus, these studies show that gypsum-slotting can increase water infiltration and availability, and improve soil aeration in sodic clay soils. Future research should focus on ways to improve the short- and long-term effectiveness of the technique. Gypsum-slotting involves creation of macropores within slots by physical manipulation, which are simultaneously chemically stabilized by the high concentration of gypsum in the slot. A combination of physical, chemical and biological techniques could be used to increase the short-term and long-term effectiveness of gypsum-slotting. Thus, the use of wider slots and high rates of gypsum would increase the short-term effectiveness of slotting. The long-term biological stabilization could be achieved by organic matter incorporation and preferential root growth in the slot. Soil pit observations showed that the swelling of the undisturbed sodic clay soil between slots during wetting could reduce slot width, leading to reductions in slot porosity and permeability (Blackwell et al., 1991). However, during drying cycles, the shrinkage of the undisturbed soils between slots leads to increased crack formation, which is the major factor influencing infiltration in swelling clay soils (Table 3).

The above studies on gypsum-slotting also illustrate the need to monitor the soil physical factors directly affecting crop yield throughout the cropping season, in evaluating the effectiveness of tillage practices. Such measurements help to relate and explain the yield responses to the tillage practice under the agronomic

Table 5. Cereal crop yields in non-ameliorated and gypsum-slotted sodic clay soils

Site	Soil	Crop/year	Non-slotted	Gypsum-slotted (0.6 m apart, 8 t ha^{-1} gypsum)	Gypsum-slotted (1.2m apart, 4 t ha^{-1} gypsum)
			Yield (t ha^{-1})		
1. Small experimental plots	Transitional red-brown earth	Wheat (1984)	5.4	8.4	
2. Large experimental plots	Transitional red-brown earth	Wheat (1986)	4.0	5.2	4.7
		Maize (1988/89)	7.0	11.5	8.1
		Maize (1989/90)	5.4	7.5	6.0
3. Farm trials	Grey clay	Wheat (1987)	3.2	4.8	
	Transitional red-brown earth	Wheat (1987)	4.1	5.7	

(From Jayawardane et al., 1991.)

management and climatic regime at the experimental site (Letey, 1985) and also provide a basis to extrapolate to other regimes. Monitoring the soil water content profiles was used as a simple field technique, to assess the soil aeration limitations in the root zone during the winter wheat crop (Jayawardane et. al., 1987). It could also be used to assess other soil physical limitations such as water deficiency or excessive soil strength (Jayawardane and Blackwell, 1990) which reduce crop yields.

V. Lime-Slotting to Improve Subsoil Acidity

Acid soils occur extensively in the high rainfall areas around the world and include over 40% of the soils in the tropics (Sanchez and Salinas, 1981). Soil acidity may occur naturally or be induced by agronomic management practices. Acidity of surface soil layers can be corrected through incorporation of lime by tillage. However surface incorporation is ineffective in correcting subsoil acidity, due to low solubility and mobility of lime. Commercial methods have not been developed for uniform mixing of lime with subsoil (McCray and Sumner, 1990), required for correcting subsoil acidity.

Subsoil acidity can affect crop growth and production in several ways. Crop losses could occur due to (a) poor root growth and function caused by aluminum, hydrogen, manganese and other metal ion toxicities, (b) deficiencies in essential nutrients for crop growth such as calcium, magnesium, phosphorus, molybdenum and iron, and (c) reduced water availability during extended drying cycles due to restricted rooting. The relative importance of these factors varies with the soil, crop and climatic conditions. Hence, site specific ameliorative practices are needed. In addition, acid clay subsoils with low permeability could benefit from deep tillage practices used for incorporating lime, because this also improves the soil physical conditions affecting root growth and function.

To evaluate the feasibility of reducing the lime rate, Kauffman and Gardner (1978) studied the effect of vertical segmental liming of an acid, Dayton silt loam soil on the growth of wheat in a growth chamber. The effect of mixing 2.2, 6.7 and 11.2 t ha^{-1} lime with 10, 30, 60, and 100% of the soil in equally spaced vertical limed segments was examined. In unlimed soil and when only 2.2 t ha^{-1} of lime was mixed with more than 30% of the soil, the roots showed aluminum toxicity symptoms due to low soil pH. Plant water stress was increased when root growth was reduced by soil acidity. Maximum wheat growth occurred when 6.7 t ha^{-1} of lime was mixed with 30% of the soil. This shows the benefit of segmental liming with low lime rates, in increasing root growth and yields. Liming of 10% of the soil volume was not sufficient to obtain maximum yield, at any lime rate.

Pearson et al. (1973) found that in a soil with alternating limed and non-limed horizontal layers spaced 0.02 - 0.08 m apart, root growth was greater than in unlimed soil. But relative root growth was less than in a completely limed soil, except at 0.02 m spacing with 60% of the soil volume limed, where root growth was similar. Root growth was related to the total mass of subsoil which was limed.

Blackwell et al. (1988) and Kirchhof et al. (1990) proposed using lime-slotting to correct subsoil acidity. The effectiveness of lime-slotting to correct subsoil acidity depends on uniformly mixing lime and soil throughout the slot. Blackwell et al. (1991) found that rotary slotting provided uniform mixing of gypsum to depth, while in soils which were ripped most of the gypsum remained in the surface layers.

Figure 4 shows soil pH changes in an acid soil caused by lime incorporation through slotting to 0.80 m depth with 20 t ha^{-1} lime, compared with delve plowing to 0.80 m depth with 20 t ha^{-1} lime, cross-ripping to 1.00 m depth with 15. t ha^{-1} lime and 10 t ha^{-1} gypsum. While slotting provided a uniform change in soil pH to the full depth of the slot, the other deep tillage practices only changed pH in the surface soil layers.

Kirchhof et al. (1991b) examined vine root growth in a lime-slotted acid clay soil, where the surface soil was also limed and mounded around the base of the vines, in a commercial vineyard. While the relative soil volumes of the slot, mound and undisturbed soil were 8, 36 and 56%, respectively, the relative amounts of roots were 30, 50 and 20% respectively. The increase in root growth in the ameliorated soil resulted in a yield of 8.2 t ha^{-1} in this vineyard, compared to 4.8 t ha^{-1} before soil amelioration.

The effect of lime-slotting on root growth, water extraction during a prolonged drying cycle and yield of a medic crop was examined in a lysimeter study, using large soil cores (1.2 m deep, 0.75 m diameter). In the non-ameliorated soil, root growth was confined to the surface 0.10 m depth. In the lime-slotted soil, with or without surface gypsum, root growth within the 0.15 cm wide slot occurred to the full depth of the slot (Figure 5), leading to greater water availability during the drying cycle, from the slot and the adjacent undisturbed soil. Crop growth was severely reduced in the non-ameliorated soil during a prolonged drying cycle. The yield increases over the non-ameliorated soil were 5-fold for lime-slotting, 8-fold for lime-slotting with surface gypsum applications and 10-fold where the full soil volume was limed.

Implements for injecting lime to depth using a deep limer (Anderson and Hendrick., 1983) and for creating a narrow open trench (0.07 m wide) into which limed topsoil can be added, have been developed. Farina and Channon (1988) found that in an acid soil injecting lime to 0.70 m depth using a deep limer or adding limed topsoil to 0.70 m depth in a 0.07 m wide trench (using a modified subsoiler) resulted in yield increases of a maize crop.

Sumner (1970) and Reeve and Sumner (1972) found that surface applied gypsum moved through a Natal Oxisol and reduced exchangeable Al in the subsoil. The beneficial effects of surface gypsum applications in gradually alleviating subsoil acidity and progressively increasing crop growth has been reported by several other workers. These studies are reviewed in detail by Shainberg et al. (1989) and McCray and Sumner (1990). Sumner et al. (1986) compared surface gypsum application with lime incorporation to depth. They

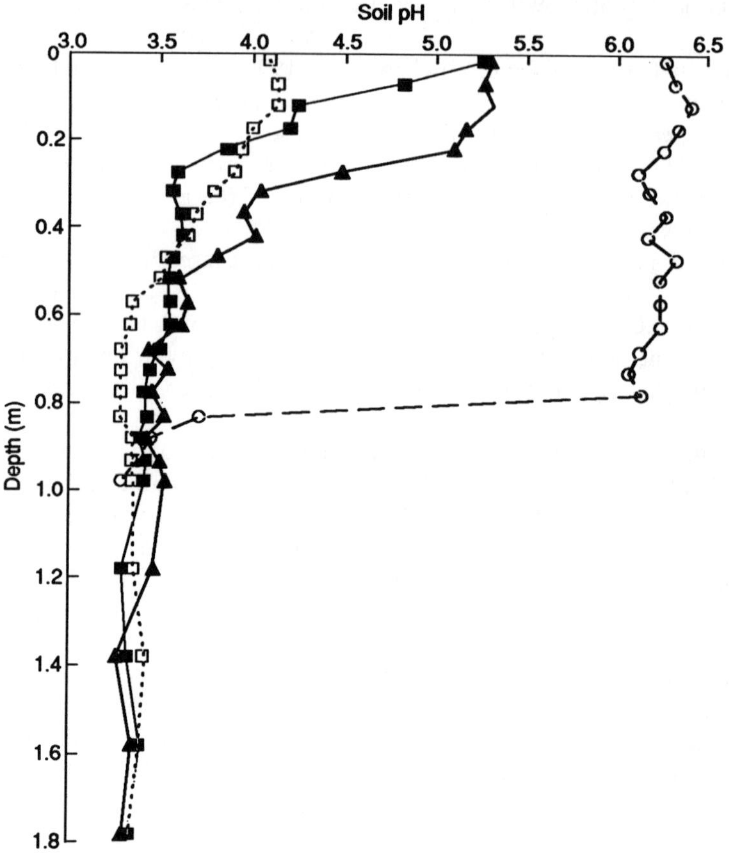

Figure 4. Soil pH in an acid soil (□), and due to gypsum slotting (○), deep ripping with lime added (▲) and delve plowing with lime added (■).

found that complete mixing of lime and subsoil, increased water extraction to the full depth of lime incorporation and increased crop yield significantly from the first to fourth year. Surface gypsum application led to marked increases in yield from the second to fourth year, mainly due to increased water extraction from the surface 0.30 m.

In acid clay soils of low permeability, lime-slotting combined with surface gypsum application is likely to lead to faster movement of gypsum to depth and quicker amelioration of subsoil acidity than surface gypsum applications alone. Furthermore, in slowly permeable acid clay soils, lime/gypsum slotting will also lead to improvement in the soil physical properties which limit crop growth, such as water infiltration and storage, soil strength (Kirchhof et al., 1990) and soil aeration.

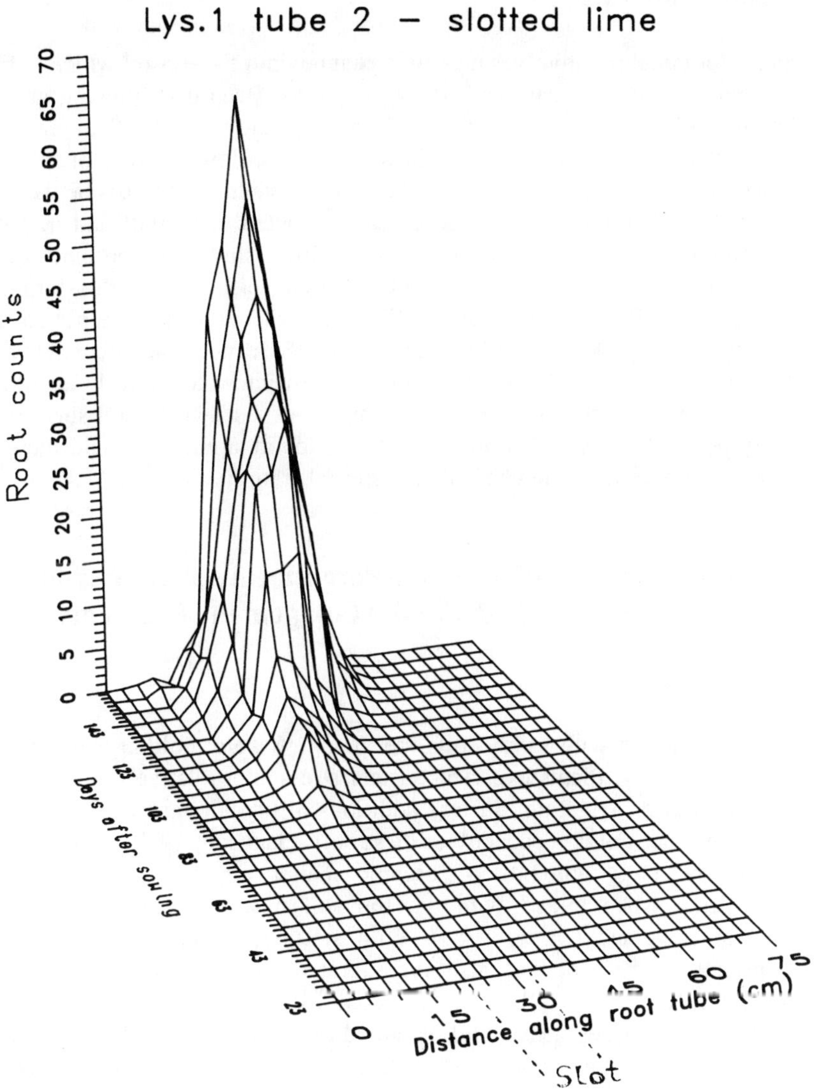

Figure 5. Changes in medic root count, during a growing season in a traverse across a slotted acid soil at 0.60 depth. (Dr. H.D. Barrs, personal communication.)

VI. Slot Mulching

Vertical slot mulching involves making a channel into the subsoil which is filled with organic residues (Spain and McCune, 1956). Deep disc plowing and slot mulching with slots 0.22 m wide, 0.60 m deep and spaced 2.0 m markedly increased the rate and depth of water infiltration on a slowly permeable Pullman clay loam (Hauser and Taylor, 1964). The lack of yield differences between the ameliorated and non-ameliorated soils was attributed to rainfall and irrigation water being held on all plots, allowing it to infiltrate slowly and become equally available during dry periods. Subsequent studies have shown that the slot mulch system is only effective in increasing infiltration, where the mulch is not covered at the surface by soil during tillage operations. Slot mulching increases water infiltration into frozen soil (Saxton, 1980), by allowing water to infiltrate into the non-frozen lower depths, through the slot mulch exposed at the soil surface. On a slowly permeable vertisol, Rao et al. (1977) found significant yield increases of sorghum, on vertical mulched plots, especially in dry years.

VII. Future Research Needs on Potential Applications of Slotting in other Soil and Land Management Problems

A. Sandy Soils

The main problems with sandy soils are their low water holding capacity and nutrient availability. Plants experience water stress after relatively short periods without rainfall or irrigation. Surface additions of organic matter can markedly increase the water holding capacity of these soils (Jamison, 1953; Unger, 1975). However, the large amounts of organic matter required and their rapid decomposition when mixed with the top soil make this method generally impractical (Eck and Unger, 1985). With slotting, organic matter could be concentrated in the slot and also incorporated to depth, where lower soil temperatures could reduce the rate of decomposition. Any nutrients or micronutrients which are deficient could be added as an amendment during slotting, or mixed into the organic matter. The organic matter could be imported, grown in-situ or available as stubble from the previous crop. The crop residues could be mulched and blown into the slot during slotting. A combined operation with the rotary slotter and a mulcher, similar to that used in vertical slot mulching, could be used. Although increased depth of rooting will occur more rapidly in the slot, water could be extracted from the adjacent undisturbed soil during extended drying cycles, as in the case of lime-slotting of acid soils.

B. Deep Placement of Nutrients and Micronutrients

Slotting could be used to place nutrients and micronutrients to depth to meet specific nutrient deficiencies, possibly in combination with other amendments such as lime, gypsum, sewage sludge and organic residues. Studies on effects of subsoil fertilization on plant responses have shown conflicting results. Several studies have shown the benefits of deep placement of fertilizers in increasing crop production (McEwan and Johnston, 1979; Patrick et al., 1959), while others have shown no responses. Sumner and Boswell (1981) suggested that the lack of responses to deep fertilizer placement in some experiments could be due to the failure to establish, at the onset, a definite need for deep fertilizer application to overcome nutrient deficiencies in the subsoil.

Fenn et al. (1990) showed that in a calcareous soil application of acid and $ZnSO_4$ in a narrow trench increased availability of Zn. Acidification of less than 1% of the root zone of pecan trees significantly increased Zn uptake and maintained elevated levels of Zn in leaves for 9 years. This technique could thus replace the more costly foliar spray method for applying Zn. Slotting with acid or $ZnSO_4$ could increase Zn availability at soil depths below the cultivation layer.

Slotting could be an effective tillage technique to bring up minerals from depth to the topsoil, where they can have a beneficial effect. For instance, McAndrew and Malhi (1990) found that deep plowing a solonetzic soil increased the calcium content and pH of upper layers leading to higher crop yields. Loveday (1984) suggested improving sodic soils underlain by calcium mineral rich layers through mixing the two layers, combined with addition of acid or acid forming amendments to release the calcium from such minerals. Use of slotting to mix soil layers should be tested on these soils. The use of slotting for deep placement of immobile nutrients such as a phosphorus and some micronutrients in soils of high or low pH, also needs to be further researched.

C. Erosion Control and Water Conservation in Dryland Agriculture

A common problem in dryland cropping of clay soils is the low infiltration rates. This leads to reduced water storage resulting in low crop productivity. This also leads to excess runoff and erosion, causing soil degradation. While soil loosening can increase infiltration, it can make the soil more erodible. Conservation tillage systems which reduce the frequency and intensity of soil manipulative operations are being increasingly adopted to reduce soil erosion. However, conservation tillage may suppress yields, especially on heavy textured soils with poor internal drainage, or on soils prone to compaction by vehicular traffic (Logan et al., 1991). Yushkevich et al. (1982) showed that the reduced water intake of minimum tilled soils could be overcome by slotting.

Gypsum/lime slotting could also be used to increase infiltration on sodic and acid clay soils used in dryland agriculture. The increased water depletion by crops in sodic and acid clay soils during drying cycles will also lead to increased intake during rainstorms (Table 3), thus reducing runoff (Blackwell et al., 1991) and erosion. Saxton (1980) found that runoff and erosion from winter rains and snowmelt on frozen soils could be reduced with slot mulch tillage, with slots located on the contour spaced 2-4 m apart. Further field research is necessary on the effectiveness of combining conservation tillage practices with soil slotting, to conserve water and soil.

D. Controlling Dryland Salinity

Dryland salinity is an important problem in southern Australia and other dry regions around the world. It is caused by the removal of deep rooted trees with high evaporative demand through the year and their replacement with shallow rooted annual crops or pasture. This leads to increased runoff, subsurface flows in perched aquifers and deep percolation which raises the ground watertable in low-lying areas (Davidson and Bell, 1989). This results in formation of saline seepage areas in the lower aspects of the landscape. Reforestation of farmlands is often not economically viable. Therefore to overcome this problem, soil and crop management systems which not only increase infiltration in the upper slopes but also fully utilize this water through evapotranspiration to minimize deep percolation to low lying areas have to be developed.

In farmland, where subsurface flow in shallow perched aquifers is the main cause of salinization, widely spaced (50-100 m), deep slots located on the contour could be used to intercept subsurface flow in perched aquifers and divert it to deeper soil layers. Deep rooted perennial crops or tree crops or fodder crops planted in this narrow slotted area could be used to remove this water by evapotranspiration through the year. This represents a combination of slotting with alley cropping (Lal 1991), to overcome the problem of excessive surface and subsurface runoff on slopes, causing saline seepage areas. Slots refilled with soil enriched with gypsum or lime could be used on sodic and acid soils, respectively.

In addition, the strips between the planted tree lines, which are used for cropping, could be lime/gypsum-slotted to shallow depths, to increase infiltration and water use during dry periods (Table 3) and thereby reduce runoff. However, the slot configuration should be carefully designed to match infiltration to the crop evapotranspiration, thereby preventing deep percolation. It is emphasized that the use of the slotting technique should be considered in combination with other management techniques, based on an understanding of the hydrology of the whole macro-catchment area which is causing the problem of dryland salinity.

In sandy soils, slotting with sewage sludge or organic matter could increase water holding capacity, depth of rooting and crop water use, leading to reduced deep percolation losses. Field research studies are needed to evaluate the use of slotting, suitably modified to meet the catchment hydrological characteristics, in overcoming the problems of dryland salinity.

E. Increasing Mole Stability in Sodic Clay Soils

Mole drainage is used in Europe to improve drainage of non-sodic clay soils. In an experiment on mole drainage on a sodic clay soil, the mole channel was partially blocked by collapse of the mole within one year (Muirhead *et al.*, 1992). When a shallow gypsum salt was positioned above the mole channel, the mole channel showed very little signs of collapse. This appears to be due to increased mole wall stability, through replacement of sodium by calcium on the surfaces of clay particles. This will lead to the mole channel lasting for a longer period of time. This technique needs to be further tested.

VIII. Slotting for Land Treatment of Urban Waste

Safe disposal of sewage sludge and other urban waste in a major problem in large cities around the world. The cost of urban waste disposal also increases with the expansion of the suburbs around the cities and as dedicated land dump sites get filled. Increasing community concern on the potential damage to environment through sewage disposal has led to demands and legislature for increasing secondary treatment of sewage. This has led to production of cleaner liquid effluent but more sludge for disposal.

Application of sludge to agricultural lands close to cities has been increasing, as it appears to be a cheap option for safe disposal. However, uncontrolled land disposal of sludge could result in contamination of land, water and food chains. Hence, in evaluating the potential for increasing sludge application on agricultural lands, the benefits and potential hazards need to be carefully considered.

Previous studies have shown that agricultural crop production can be increased by sludge applications and incorporation in surface soil layers. This is due to sludge being a low grade fertilizer - supplying nitrogen, phosphorus and minor plant nutrients, depending on its quality (Sommers, 1977). However, associated with the low nutrient concentrations is the high transportation costs of sludge (per unit weight of nutrients) when compared to inorganic fertilizers. Sludge application has also been shown to improve soil physical properties, such as water retention capacity and water stability of aggregates (Metzger and Yaron, 1987). The high organic matter content could also be beneficial in ameliorating acid soils (Evans and Kamprath, 1970; Hue et al., 1986) and hence reduce the amount of lime required.

In addition to providing these physical and chemical soil improvements, sewage sludge-slotting has several advantages over sludge disposal by incorporation into surface soil layers of agricultural lands (Kirchhof et al., 1991a). Where the sewage sludge contains nitrogen, phosphorus or heavy metal contaminants, the rate of surface application needs to be carefully controlled, due to the risk of pollution through erosion and runoff into streams. With sludge-slotting much higher rates of sludge could be applied to agricultural land, since only a small amount of the sludge gets incorporated into the surface layers. The chance of leaching nitrogen and heavy metals from the slots into the groundwater is reduced in heavy clay soils of low permeability. Losses of phosphorus through surface runoff or leaching will be minimal. Therefore, the nitrogen and phosphorus could be used by crops over several seasons. The capacity to dispose of large quantities of sewage sludge through sludge-slotting will reduce the costs of transport and disposal of sludge on the limited area of agricultural land available close to cities. Use of higher sludge rates also increases the benefits in terms of soil improvement for crop production from each application. For slots 0.8 m deep, spaced 1 m apart in a cross-hatched pattern, sludge application rates of 350 t h^{-1} were achieved (Kirchhof et al., 1991a). In a preliminary field experiment, sludge-slotting on an acid clay soil led to deeper rooting and higher yields of a barley crop compared to surface application at the same rate. Sludge-slotted lands will produce less odor problems than where the sludge is incorporated into surface soil layers. Composted urban garbage and other urban and rural waste may also be disposed of by slotting, especially in lands allocated to forestry and parks.

Where domestic sludge is mixed with industrial sludge, the risk of heavy metal contamination needs to be carefully monitored (Tiller, 1989). Adding lime to sewage sludge used for slotting could immobilize heavy metals in the sludge. However, with high levels of heavy metal contamination, cropping may need to be restricted to non-consumable crops such as forestry and floriculture. Further research is needed to investigate the short-term and long-term release of heavy metals immobilized by limed sludge-slotting on acid clay soils. Sikora and Radecke (1987) applied limed-sludge in wide sludge trenches located in farmers fields, at application rates 10 times higher than recommended for surface application (10 t ha^{-1}). The maximum heavy metal contents in crops grown on the limed-sludge sites were lower than the maximum recorded in crops amended with commercial fertilizer. While poor surface soil conditions created in excavating wide trenches adversely affected plant growth, this will not be a problem in sludge-slotted soils, where there is minimal disturbance.

The sludge could also be combined with other ameliorants used for slotting to improve soils with adverse subsoil conditions. Jayawardane and Blackwell (1985) and Jayawardane et al. (1987) showed that sodic clay soils with poor infiltration and internal drainage problems could be ameliorated by gypsum-slotting. A mixture of sludge and gypsum could further improve the physical stability of the macropores, thereby increasing the initial and long-term effectiveness of gypsum-slotting. It will also encourage earthworm activity on these soils, thereby increasing the macroporosity of slots (Blackwell and Blackwell 1989). Blackwell et al. (1988), Kirchhof et al. (1991a) and Jayawardane et al. (1991) have shown the benefits of lime-slotting of acid soils, leading to increased root growth and crop production. Using a mixture of sludge and lime on acid soils will increase the long-term stability of the slots and immobilize heavy metals in the sludge. Sludge-slotting combined with supplementary addition of nutrients and micronutrients could increase the water and nutrient retention capacity of sandy soils. However, in sandy soils the potential risk of leaching of nitrogen and heavy metal from the sludge slots to underground water sources should be carefully monitored.

Thus, sludge-slotting could provide an environmentally safer method for disposal of sewage sludge and other urban and rural waste. In addition, it could be used in combination with other ameliorants to ameliorate problem soil with adverse subsoil conditions, which are difficult to ameliorate with conventional ameliorative techniques. These problem soils include sodic, acid and sandy soils, which have very low cropping potential when they are not ameliorated. Use of sludge-slotting on these problem soils could therefore maximize the monetary returns to sewage-sludge disposal on agricultural lands. This will increase the economic viability of this method of sludge disposal.

IX. Slot Protection from Recompaction during Trafficking

Blackwell et al. (1989b) compared the vertical stresses during wheeling across slots and ripping troughs, using stress transducers (Horn 1980, 1986) buried in the slots, ripping troughs and adjacent undisturbed soil. They found that the slots were protected from compressive stresses during wheeling due to bridging of the wheel "contact patch" across the slot, and the stronger uncultivated soil adjacent to the slots supporting much of the wheel load. In contrast, stresses within the ripping trough were similar to those outside the ripping trough, and greater than in the slots. The magnitude of the changes in bulk density, air-filled porosity and intrinsic permeability were greater in the ripping troughs compared to the slots, after 1 and 4 passes of the tractor wheel. While there is some protection of the macro-pore structure of the slots during trafficking, damage could be caused by excessive wheeling, especially in wet soil where the strength of the undisturbed soil is low.

The extent of protection of the loosened soil within the slot depends on the slot width, direction of wheeling and the number of traverses (Blackwell et al., 1989a). For both transverse wheeling (i.e., wheeling at right angles to the direction of the slot) and longitudinal wheeling (i.e., wheeling in the direction of the slot with the center of the travel path coincident with the center of the slot), the vertical stresses in the slot were decreased with reduction in slot width (Figure 6). During transverse wheeling, an increase in slot width results only in slight increase in stresses as long as a critical slot width was not exceeded. If, on the other hand, the slots are wheeled longitudinally the slotted soil is less protected, because of the higher proportion of the slot to the undisturbed soil area beneath the wheel, for any given slot width.

In the undisturbed soil, the highest stresses were measured during the first wheeling. However, the formation of a more compacted soil volume near the soil surface during consecutive wheeling led to a more pronounced stress attenuation, with smaller vertical stress values measured at any given depth (Blackwell et al., 1989a). Inside the slot the trend was reversed. With repeated traversing, a

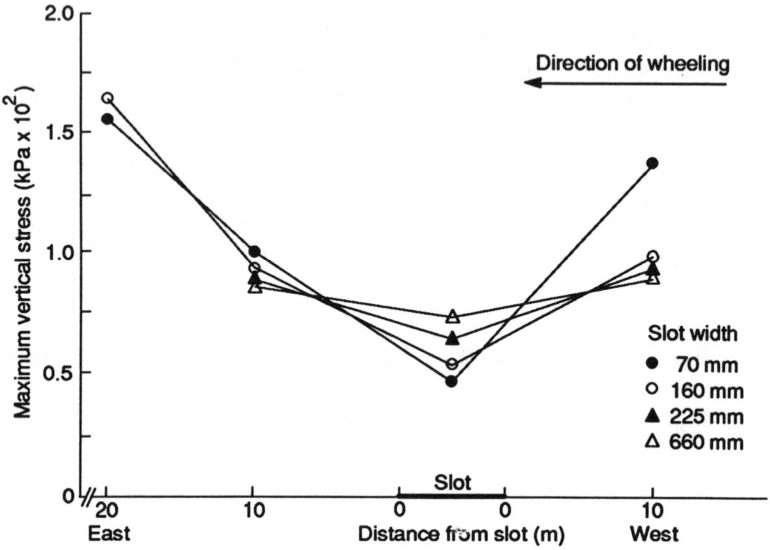

Figure 6. Effects of slot width on maximum vertical stress at 0.20 m depth during transverse wheeling, within and between slots. (From Blackwell et al., 1989a.)

recently slotted soil became more compacted as shown by the increasing stress values.

Mean vertical stress values at all depths inside a recently formed slot were similar, regardless of slot width. This can be explained by the absence of cementing of the slotted material with the adjacent undisturbed soil, especially immediately after the preparation of the slots, as the friction between the very loose slotted soil and the smoothed vertical slot wall is small. Thus, any applied stress results in compaction of the entire slotted soil volume. Consequently, the applied stresses will be attenuated by the undisturbed soil below the slot (Terzaghi and Jelinek, 1954; Horn et al., 1987).

Using simulation experiments and analytical studies, Kirby and Blackwell (1989) showed that dry, V-shaped slots with rougher sidewalls provided greater protection for slot contents during wheeling, than wet, vertical slots with smooth walls.

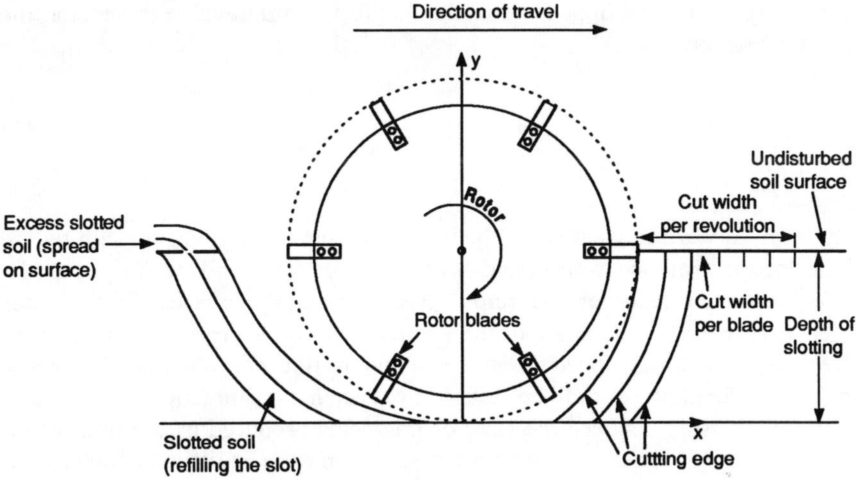

Figure 7. Schematic diagram of slot formation by a rotary slotter.

X. The Process of Slot Formation by a Rotary Slotter and Fragment Size Distribution within Slots

A diagrammatic sketch of the process of slot formation by rotary slotting and refilling of the slot is given in Figure 7. The thickness of cut (T_c mm) is dependent on the tractor speed (S_t km h^{-1}), rotor speed (S_r rev min^{-1}) and the number of blades per rotor (N), and can be determined using Equation [1].

$$T_c = \frac{10^6 S_t}{60 S_r N} \qquad [1]$$

The path of the cutting arc of the rotor for different thickness of cut (produced by any combination of tractor and rotor speeds) can be plotted as follows, on the x and y coordinates shown in Figure 7. When the tractor is not moving and the thickness of cut is zero, the cutting arc is part of a circle. The displacement of this cutting arc from the circular path as the tractor speed increases is described by the correction term included in Equation [2]:

$$x = (2ry - y^2)^{1/2} + \frac{N}{4} T_c \left(\frac{r - y}{r}\right) \qquad [2]$$

where r is the radius of the rotor (measured to the tip of the rotor blades). If the rotors are rotated in the opposite direction to the forward travel of the tractor, the equation becomes:

$$x = (2ry - y^2)^{1/2} + \frac{N}{4} T_c \frac{y}{r} \qquad [3]$$

Due to the forward movement of the tractor during the slotting, the cutting arc of the forward rotating rotors (and hence the slot face) will be steeper as the tractor speed increases or the rotor speed decreases. However, if the rotor direction is reversed, the angle of the slot face will be lowered as tractor speed increases or the rotor speed decreases. Field testing showed that the power requirement for slotting could be reduced by reversing the direction of the rotors.

The relationship between the sizes of fragments produced by slotting with a prototype rotary slotter and the tractor speed and the moisture content of a transitional red-brown earth soil was examined (Jayawardane et al., 1988). In this poorly-structured soil, the three rotors spaced 1.2 m apart, working in narrow (0.12 m wide) parallel bands, produced well-defined 0.40 m deep slots. The slots were filled with platy-structured fragments, formed by the slicing action of the rotors on the slot face and smaller blocky fragments produced by subsequent vigorous mixing within the rotor shrouds.

The minor axis (third largest dimension) of the platy fragments is determined by the thickness of cut produced by 2 consecutive rotor blades (Figure 7), which is related to the forward speed of the tractor and the rotor speed. The sizes of the other two dimensions are determined largely by the nature and extent of secondary shattering of the fragments, although the width of the rotor blades is a contributory factor. The intermediate axis (second largest dimension) determines the passage of the fragments through the sieve openings. The geometric mean diameter of these fragments and the percentage of different sized fragments were closely related to the thickness of cut (Table 6). This shows that the extent of secondary shattering is directly related to the thickness of cut. More than 90 and 70% of the fragments had diameters < 25 mm and 11 mm, respectively, at the tractor speeds (0.36-0.70 km h^{-1}) and the rotor speed (94 rpm) used in the experiment. Within the slots, the percentages of the smaller-sized fragments increased with depth, possibly owing to interference in the deposition of the larger sized fragments by the rotors.

The deposition of the soil fragments occurs in the slot behind the rotor and mostly in the space below the rear half of the moving rotors. The thickness of cut per revolution of the rotors within the slot is equal to the thickness of fill within the slot, since the rotor and blades occupy a constant space within the slot. The decreased bulk density, and hence the increase in volume of the fill material, causes the excess soil to be deposited on top of the slot and also spread on the surface of the undisturbed soil. However, since the soil fragments formed by slotting are chiefly less than 25 mm in size (Table 6) a seedbed can be easily

Table 6. The mean percentage fragnent size distribution within the slot and the thickness of cut, corresponding to different tractor speeds

Tractor speed	Thickness of cut	$FS_1 > 25$	FS_2 11-25	FS_3 4.8-11	FS_4 2-4.8	$FS_2 < 2$	GMD*
($Km\ h^{-1}$)	(mm)	Percentage of fragments of different sizes, mm					
0.36	10.64	1.8^{c}	19.1^{c}	26.0^{b}	17.9^{b}	35.2^{a}	0.56
0.50	14.77	3.0^{c}	23.9^{abc}	22.2^{ab}	15.4^{b}	35.5^{a}	0.60
0.60	17.73	7.0^{b}	26.1^{ab}	21.9^{ab}	14.8^{b}	30.3^{a}	0.91
0.70	20.57	10.4^{a}	26.4^{a}	19.1^{a}	13.4^{a}	30.6^{a}	0.95
Mean		4.5	23.3	23.0	15.7	33.5	0.69

Means with superscripts containing common letters are not significantly different ($P < 0.05$, Duncan Multiple Range Tests).
* Geometric mean diameter (GMD) of the fragments was calculated using

$$GMD = Antilog \frac{\sum Log\ d_1\ x\ w_i}{\sum w_i}$$

where d_i and w_i are the lower diameters and weights, respectively, of each fragment size class, and the value of d_i for FS_5 fragments was taken as 0.01 mm.
(From Jayawardane et al., 1988.)

formed with a light working of the surface soil. This contrasts with deep ripped soil, where large clods which require further tillage operations to break them down, are often formed.

The fragment size distribution was only slightly affected by the water content at the time of slotting (Jayawardane et al., 1988), over the water contents close to the lower plastic limit of the soil. This is in contrast to the marked effects of soil water content on the soil failure during operation of rigid tines. Fragmentation of soil by deep ripping tines is more effective at water contents less than the lower plastic limit. Spoor and Godwin (1978) showed that deep ripping close to the lower plastic limit can lead to the tine forcing a slit into the soil by lateral compression, if the tine is working below its critical depth. Thus in contrast to ripping, slotting can be carried out over a wider range of water contents, thereby increasing the time available for this tillage operation on farms.

While rotary slotting can be done over a wide range of soil water contents, it is desirable to slot at high water contents, provided that excessive deformation of slot fragments does not occur. Slotting at higher water contents reduces the power requirements for cutting the soil and allows faster tractor speeds and increases the rate of slotting operations. Faster tractor speeds also produce thicker cuts and coarser slot fragments, leading to increased macro-porosity of the slots.

XI. Design and Construction of Rotary Slotting Implements

The configuration of the slots required to provide optimum conditions for plant growth varies according to the soils, crops, climates, irrigation methods and other agronomic practices adopted. Hence a requirement in the design of the implement should be that it is sufficiently versatile to provide changes in the spacing, width and depth of the slots with simple adjustments. Rotary tillage appears to meet the design requirements. It also has the advantage of low draft requirements. Rotary tillage can be carried out using disc shaped rotors fixed onto a rotating shaft and adjusted to give the required slot spacing. On the periphery of these rotors, the rotary blades of any required width are bolted.

Several prototype rotary slotters have been built. The first prototype consists of a single rotor with a shroud, capable of digging slots 0.15 m wide and 0.40 m deep. The single rotor slotting mechanism and shroud is attached to the back of a tractor and lowered to work at the required depth. The rotors are driven through a differential connected to the tractor power-take-off. The soil amendment is applied in a separate operation before slotting, in bands on the soil surface at the required slot spacing and rate of application. This could be done using existing gypsum/lime spreaders modified to apply the amendment in a band. This implement can be used on small farms where slotting to shallow depths is required.

The second prototype (Figure 8) is described in detail by Blackwell et al. (1990). It was designed to carry out gypsum-slotting of sodic clay soils to 0.4 m depth. The rotor assembly consists of 3 rotors to cut slots at a maximum spacing

(a)

(b)

Figure 8. Rotary slotters (a) Prototype (b) Prototype 5.

of 1.2 m. Gypsum is carried in a hopper. The problem of bridging and compaction was overcome by the use of two augers with short flighted sections to move gypsum to the hopper openings, and a series of pegs to agitate the gypsum in the bottom of the hopper. Different gypsum application rates are obtained by adjusting the auger speed and the size of the gypsum hopper openings (Blackwell et al., 1990). The gypsum is added and mixed with the soil within the shroud. On a dry hard sodic clay soil this implement was used with a 120 kW tractor at tractor speeds of up to 0.5 km hr^{-1}. On wetter soils, faster working speeds could be used. Prototype 2 is a trailed implement with four wheels. Subsequently, it was modified into a more maneuverable implement (prototype 3) with two wheels. Prototype 4 is a larger machine built for lime-slotting and sludge-slotting to a depth of 0.8 m (Kirchhof et al., 1991). Prototype 5 was a smaller machine built to slot between vine rows in established vineyards (Figure 8).

Several non-rotary type slotting or trenching machines have been built (Kiktev and Mitrofanov, 1979; Hyde et al., 1986; Farina and Channon, 1988). However, no information is available on effects of these implements on the slot aggregate sizes in different soils and working conditions. Trench digging machines with cutting chains have sometimes been used in experimental studies.

The best choice of type of slotting implement to use varies according to the crop, the nature and extent of soil problem to be solved and economic factors. Further research is required to improve the engineering efficiency of these prototypes in developing these implements for wider commercial use.

XII. Choosing the Optimum Slot Configuration

The successful field application of the slotting technique requires a careful identification and assessment of the soil factors which limit crop production in the non-ameliorated soil under existing climatic conditions (Letey, 1985). The extent to which the soil amelioration technique is likely to overcome any root zone limitations during the growing season also needs to be considered. Therefore, the optimum slot configuration and type and amount of ameliorant to be added varies widely according to the soil, crop, climate and agronomic management practices adopted on the site. The availability and price of the ameliorants and other economic factors also need to be considered.

Modeling may be used to predict the optimum slot configuration for specific sites and expectations of weather patterns. Models for predicting the two-dimensional infiltration into vertical slot mulches have been developed (Swartzendruber, 1960; Saxton and Campbell, 1983; Redinger et al., 1984). However, more complex numerical models need to be developed for evaluating the two-dimensional water redistribution and water extraction patterns in slotted soils, which take into account hysteresis, different rates of root growth in slotted and non-slotted areas and the changes in hydraulic properties of the soil with leaching of gypsum and other amendments with time. In swelling soils, crack-filling (of the cracks formed during drying cycles) provides the main component of infiltration. In these soils, other factors such as the effect of gypsum-slotting on crack formation, on drying of surface soil layers, and on relief of lateral pressures through swelling of the undisturbed soil into the slot need to be considered.

Where the main limitation to crop production in non-ameliorated soil is a shortage of water during prolonged dry periods, simpler models may be used. For instance, in the lime-slotted acid soil the root distribution is largely confined to the slot. The availability of water during dry periods is determined by the plant water available in the slot and the rate of flow of water from the undisturbed soil to the dry slot. This is largely determined by the hydraulic properties of the undisturbed soil.

Carefully designed field experiments could also be used to determine the optimum slot configuration. Measurements of the changes in the two-dimensional soil water content profiles through the season during such experiments could be used to test the model predictions and to provide a basis for extrapolating the experimental results to other climatic regimes.

XIII. Summary

Slotting creates narrow parallel bands of loosened soil into which an ameliorant or a combination of ameliorants is added leaving the rest of the soil profile undisturbed. It is an effective technique for long-term amelioration of sodic, acid and other problem soils.

Slotting of soils with clay pans and impermeable layers, combined with gypsum as an ameliorant in sodic soils, increases root growth, and water infiltration, availability and extraction. In addition, internal soil drainage and aeration is improved, both within and between slots, leading to increased crop production. Lime-slotting of acid soils uniformly changes soil pH to depth within the slot. It also improves the soil physical factors limiting root growth and function in acid clay soils, leading to improved root growth and crop production.

Slotting is a long-term land development technique, as slot recompaction during trafficking is minimized by bridging of the tractor wheels across the slot onto the undisturbed stronger soil.

Sludge-slotting is a useful technique for land treatment of urban waste on agricultural lands, which could minimize the risk of environmental pollution. In addition to the benefits of improving soil physical properties and supplying nitrogen and phosphorus for crop growth, sludge-slotting allows the sludge to be partly or fully buried in the ground. This could reduce or eliminate the risk of pollution of surface water resources with nitrogen, phosphorus or heavy metals through runoff. In clay soils of low permeability the loss of these nutrients through leaching will be minimal. Future research is necessary to examine these chemical processes in detail and to examine the role of adding lime to sludge-slots to immobilize heavy metals.

The optimum slot configuration and the type and amounts of ameliorants to be added varies widely according to the soil, crop, climate and agronomic management practices to be adopted on the site. The successful application of tillage techniques requires a careful identification and assessment of the soil factors which limit crop production in the non-ameliorated soil under the existing climatic conditions (Letey, 1985) and the extent to which the techniques are likely to overcome such root zone limitations during the cropping season (Eck and Unger, 1985). Monitoring the soil water content profiles in non-ameliorated and gypsum-slotted sodic clay soil, throughout a wheat cropping season with a wet winter, provided an assessment of periods when soil aeration was limiting in the non-ameliorated soil, in relation to the growth stages of the crop. These were related to reductions in wheat yield components in the non-ameliorated soil, compared to the slotted soil. Monitoring the changes in soil water profiles could also be used to assess other soil physical limitations to crop growth, such as water deficiency and excessive soil strength, during critical crop growth stages.

Several prototype rotary slotters have been built and successfully field tested in field trials and farms. Rotary slotting produces distinct slots of specified width, depth and spacing, filled with platy or blocky soil fragments. It provides uniform incorporation of the amendment to depth. The size of the fragments is dependent on the thickness of cut, which is directly related to the tractor and rotor speed, but is largely independent of soil water content. This allows greater timeliness in carrying out slotting in the field. However, slotting at higher water content is preferred because it reduces power requirements and the time taken for slotting. Faster tractor speeds also increase the thickness of cut and hence increase fragment size and hence initial slot macroporosity. The need to further investigate factors which increase short-term and long-term effectiveness of slots in sodic soils, such as width of slots, size of slot aggregates, sludge or organic matter incorporation and the amount of gypsum required, is emphasized. Design and construction of improved slotting implements is a challenge for agricultural engineers.

Other potential future applications of slotting which need to be researched are (a) effects of slotting sewage-sludge, organic matter or crop residues in sandy soil (with nutrient or micronutrient addition where necessary), on soil water retention and availability, and on crop growth, (b) prevention of dryland saline seep formation in deforested lands used for agriculture, and (c) combining slotting with minimum tillage practices to reduce soil erosion and improve water conservation, while maintaining crop productivity in dryland agriculture.

References

Adams, F. 1981. Alleviating chemical toxicities: liming acid soils. p.269-304. In: G.F. Arkin and H.M. Taylor (eds.) *Modifying the root environment to reduce crop stress*. Am. Soc. Agric. Eng., Michigan.

Anderson, D.L. and J.G. Hendrick. 1983. Subsoil lime injector. *Soil Sci. Soc. Am. J.* 47:337-339.

Bentley, O.G. 1985. Soil erosion and crop productivity: A call for action. p. 1-7. In: R.F. Follett and B.A. Stewart (eds.) *Soil erosion and crop productivity*. ASA/CSSA/SSSA Publishers Madison, Wisconsin.

Blackwell, P.S. N.S. Jayawardane, T.W. Green, J.T. Wood, J. Blackwell, and H.J. Beatty. 1991. Soil macropore space of a transitional red-brown earth after either deep tillage, gypsum or both. I. Physical effects and short term changes. II. Chemical effects and long term changes. *Aust. J. Soil Res.* 29:123-140, 141-154.

Blackwell, J., N.S. Jayawardane, and R.K. Butler. 1990. Design, construction and preliminary testing of a slotting implement for gypsum enrichment of soil. *J. Agric. Engng. Res.* 46:81-92.

Blackwell, P.S. and Blackwell, J. 1989. The introduction of earthworms to an ameliorated irrigated duplex soil in south-eastern Australia and the influence on macropores. *Aust. J. Soil Res.* 27:807-814.

Blackwell, J., R. Horn, N.S. Jayawardane, R. White, and P.S. Blackwell. 1989a. Vertical stress distribution under tractor wheeling in a partially deep loosened Typic Paleustalf. *Soil Tillage Res.* 13:1-12.

Blackwell, P.S., N.S. Jayawardane, J. Blackwell, R. White, and R. Horn. 1989b. Evaluation of soil recompaction by transverse wheeling of tillage slots. *Soil Sci. Soc. Am.* 53:11-15.

Blackwell, J., N.S. Jayawardane, W.A. Muirhead, and R. Smart, 1988. Lime-slotting technique for amelioration of acid soils. p.15. J. Loveday (ed.) *National Soils Conference*. Aust. Soc. Soil Sci. Canberra, ACT.

Blaikie, S.J., K.B. Kelly, W.K. Mason, and F.M. Martin. 1988. Productivity of irrigated pastures following paraplowing and gypsum slotting of subsoil exposed by laser land forming – a short term study. *Aust. J. Agric. Res.* 39:837-845.

Bradford, J.M. and R.W. Blanchar. 1977. Profile modification of a Fragiudalf to increase crop production. *Soil Sci. Soc. Am. J.* 41:127-131.

Bradford, J.M. and R.W. Blanchar. 1980. The effect of profile modification of a Fragiudalf on water extraction and growth by grain sorghum. *Soil Sci. Soc. Am. J.* 44:374-378.

Cockroft, B. 1984. Improving the root environment. pp. ix-xi. In: W.A. Muirhead and E. Humphreys (eds.) *Root Zone Limitation to Crop Production on Clay Soils*. CSIRO Aust., Melbourne.

Collis-George, H. 1980. A pragmatic method to determine the parameters that characterise ponded infiltration. *Aust. J. Soil Res.* 18:111-117.

Cook, F., N.S. Jayawardane, and J. Blackwell. 1989. Effects of amelioration on infiltration characteristics of a transitional red-brown earth. *New Zealand Crop and Hort. Sci.* 17:183-188.

Davidson, S. and A. Bell. 1989. Revegetation – one answer for salt-degraded land. *Rural Research*, CSIRO, Australia. 143: 14-21.

Eck, H.V. and P.W. Unger. 1985. Soil profile modification for increasing crop production. *Adv. Soil Sci.* 1:65-100.

Eck, H.V. and H.M. Taylor. 1969. Profile modification of a slowly permeable soil. *Soil Sci. Soc. Am. Proc.* 33:779-783.

Evans, C.E. and E.J. Kamprath. 1970. Lime response as related to percent Al saturation, solution Al, and organic matter content. *Soil Sci. Soc. Am. Proc.* 34:893-896.

Farina, M.P.W. and P. Channon. 1988. Acid-subsoil amelioration: I. A comparison of several mechanical procedures. *Soil Sci. Soc. Am. J.* 52:169-175.

Fenn, L.B., H.L. Malstrom, T. Riley, and G.L. Horst. 1990. Acidification of calcareous soils improves zinc absorption of Pecan trees. *J. Am. Soc. Hort. Sci.* 1155:741-744.

Fritton, D.D. and G.W. Olson. 1972. Bulk density of a fragipan soil in natural and disturbed profiles. *Soil Sci. Soc. Am. Proc.* 36:686-689.

Hauser, V.L. and H.M. Taylor. 1964. Evaluation of deep-tillage treatments on a slowly permeable soil. *Trans. Amer. Soc. Agric. Eng.* 134-136.

Heilman, M.D. and C.L. Gonzalez. 1973. Effect of narrow trenching in Harlingen clay soil on plant growth, rooting depth, and salinity. *Agron. J.* 65:816-819.

Horn, R. 1986. Auswirkung unterschiedlicher Bodenbearbeitung auf die mechanische Belastbarkeit von Ackerboden Z. Pflanzenernahr. *Bodenk.* 149:9-18.

Horn, R. 1980. Die Ermittlung der vertikalen Druckfortpflanzung in Boden mit Hilfe con Dehnungsmessstreifen. Z. Kulturtech. *Flurbereinig.* 21:343-349.

Horn, R., N. Burger, M. Lebert, and G. Badewitz. 1987. Druckfortpflanzung unter langsam fahrenden Traktoren. Z. Kulturtech. *Flurbereinig.* 28: 94-102.

Hue, N.V., G.R. Craddock, and F. Adams. 1986. Effect of organic acids on aluminium toxicity in subsoils. *Soil Sci. Soc. Am. J.* 50:28-34.

Hyde, G.M., J.E. George, K.E. Saxton, and J.B. Simpson. 1986. A slot-mulch implement design. *Trans. Am. Soc. Agric. Eng.* 291:20-25.

Jamison, V.C. 1953. Changes in air-water relationship due to structural improvements of soils. *Soil Sci.* 76: 143-151.

Jayawardane, N.S. and J. Blackwell. 1990. Use of the neutron method in assessing changes in soil strength of undisturbed and ameliorated Transitional red-brown earths during soil drying cycles. *Aust. J. Soil Res.* 28:167-176.

Jayawardane, N.S. and E.L. Greacen. 1987. Aeration status of a duplex swelling soil determined by the neutron method. p. 37-39. In: Conf. on *Measurement of Soil and Plant Water Status*. Utah State University, July 1987.

Jayawardane, N.S. and J. Blackwell. 1986. Effects of gypsum slotting on infiltration rates and moisture storage in a swelling clay soil. *Soil Use Manage.* 2:114-118.

Jayawardane, N.S. and J. Blackwell. 1985. The effects of gypsum-enriched slots on moisture movement and aeration in an irrigated swelling clay soil. *Aust. J. Soil Res.* 23:481-492.

Jayawardane, N.S. and W.S. Meyer. 1985. Measuring air filled porosity changes in an irrigated swelling clay soil. *Aust. J. Soil Res.* 23:15-22.

Jayawardane, N.S., J. Blackwell, W.A. Muirhead, G. Kirchhof, and R. Smart. 1991. Gypsum/lime slotting - A new deep tillage technique for improved crop production on sodic/acidic soils. p. 34-43. In: *Soil tillage and agricultural sustainability.* ISTRO - 12th International Conference, University of Ibadan, Ibadan, Nigeria.

Jayawardane, N.S., P.S. Blackwell, J. Blackwell, W.A. Muirhead, C. Pender, and L. Bonaventura. 1990. Effects of soil amelioration practices on soil characteristics and water intake in a sodic clay soil. p. 72-78. In: *Soil and Land Use in the 1990's.* Aust. Soc. of Soil Sci., Victorian Branch, Tetra-Branch Conference, Bendigo, July 1990.

Jayawardane, N.S., J. Blackwell, and P.S. Blackwell. 1988. Fragment size distribution within slots created in a duplex soil by a prototype rotary slotter. *Soil Tillage Res.* 12:53-64.

Jayawardane, N.S., J. Blackwell, and M. Stapper. 1987. Effects of changes in moisture profiles of a transitional red-brown earth with surface and slotted gypsum applications on the development and yield of a wheat crop. *Aust. J. Agric. Res.* 38:239-251.

Jayawardane, N.S., W.S. Meyer, and H.D. Barrs. 1984. Moisture measurement in a swelling clay soil. *Aust. J. Soil. Res.* 22:15-22.

Kauffman, M.D. and E.H. Gardner. 1978. Segmental liming of soil and its effect on the growth of wheat. *Agron. J.* 70:331-336.

Kelly, K.B., S.J. Blaikie, and W.K. Mason, 1987. Pasture response to amelioration of exposed subsoil. I. p. 320. In: *Agronomy 1987 - Responding to change.* Proc. 4th Aust. Agron. Conf. La Trobe University, Melbourne, Victoria August 1987.

Kiktev, I.U.M. and A.S. Mitrofanov. 1979. The soil-slotting and sowing machine ShChSg-1 (Equipment). *Lesnoe Khoziactvo* 6:51-53

Kirby, J.M. and P.S. Blackwell. 1989. Design of soil slots to resist wheel track compaction. *Soil Technol.* 2:147-161.

Kirchhof, G., J. Blackwell, L. Rawlinson, R. Smart, and J. Walsh. 1991a. Use of soil slotting to incorporate sewage sludge as a soil ameliorant. *Residuals Management Conf.* AWWA/WPCF North Carolina, August 1991.

Kirchhof, G., J. Blackwell, and R.E. Smart. 1991b. Growth of vineyard roots in segmentally ameliorated acidic subsoils. p. 447-452. In: R.J. Wright, V.C. Baligar, and R.P. Murrman (eds.) Proc. 2nd Int. Symposium plant soil interactions at low pH. Kluwer Academic Publ., Dordrecht, The Netherlands.

Kirchhof, G., N.S. Jayawardane, J. Blackwell, and W.A. Muirhead. 1990. Amelioration of acidic subsoils. p. 49-55. In: *Stable Cropping Systems.* 5th Australian Soil Conservation Conference, Western Australian Department of Agriculture, Perth, March 1990.

Lal, R. 1991. Myth and scientific realities of agroforestry as a strategy for sustainable management for soils in the tropics. *Adv. Soil Sci.* 15: 91-137.

Letey, J. 1985. Relationship between soil physical properties and crop production. *Adv. Soil Sci.* 1:277-294.

Logan, T.J., R. Lal, and W.A. Dick. 1991. Tillage systems and soil properties in North America. *Soil and Tillage Res.* 20: 241-270.

Loveday, J. 1984. Amendments for reclaiming sodic soils. p. 220-227. In: S. Shainberg and J. Shalhevet (eds.). *Soil salinity under irrigation. Processes and management.*

Loveday, J., J.E. Saunt, P.M. Fleming, and W.A. Muirhead. 1970. Soil and cotton responses to tillage and ameliorant treatments in a brown clay soil. 1. Soil response and water use. *Aust. J. Exp. Agric. Anim. Husb.* 10:313-324.

McAndrew, D.W. and S.S. Malhi. 1990. Long-term effect of deep plowing solonetzic soils on chemical characteristics and crop yield. *Can. J. Soil Sci.* 70:565-570.

McCray, J.M. and M.E. Sumner. 1990. Assessing and modifying Ca and Al levels in acid subsoils. *Adv. Soil Sci.* 14:45-75.

McEwan, J. and A.E. Johnston. 1979. The effects of subsoiling and deep incorporation of P and K fertilisers on yield and nutrient uptake of barley, potatoes, wheat and sugarbeet grown in rotation. *J. Agr. Sci.* 92: 695-702.

McIntyre, D.S., J. Loveday, and C.L. Watson. 1982. Field studies of water and salt movement in an irrigated swelling clay soil. I. Infiltration during ponding. II. Profile hydrology during ponding. III. Salt movement during ponding. *Aust. J. Soil Res.* 20:81-90, 90-91, 101-105.

Metzger, L. and B. Yaron. 1987. Influence of sludge organic matter on soil physical properties. *Adv. Soil. Sci.* 7:141-161.

Meyer, W.S., H.D. Barrs, R.C.G. Smith, N.S. White, A.D. Heritage, and D.L. Short. 1985. The effect of irrigation on soil oxygen status and root and shoot growth of wheat in a clay soil. *Aust. J. Agric. Res.* 36:171-185.

Miller, D.E. 1986. Root systems in relation to stress tolerance. *HortScience* 21:963-970.

Muirhead, W.A., E. Humphreys, N.S. Jayawardane, J. Moll, and D. Foley. 1992. Watertable control using shallow subsurface drains in irrigated soils. p. 15. In: *Soil protection and productivity.* 4th National Conf. of the Aust. Soc. Soil Sci.

Patrick, W.H., Jr., L.W. Sloane, and S.A. Phillips. 1959. Response of cotton and corn to deep placement of fertiliser and deep tillage. *Soil Sci. Soc. Am. Proc.* 23:307-310.

Pearson, R.W., J. Childs, and Z.F. Lund. 1973. Uniformity of limestone mixing in acid subsoils as a factor in cotton root penetration. *Soil Sci. Soc. Am. Proc.* 37:727-732.

Polupan, N.I. 1982. Trench plowed soils of the dry Ukrainian steppe in the classification system. *Soviet Soil Sci.* 135:78-88.

Rao, M.S., V.R. Rao, M. Ramachandran, and R.C. Agnihotri. 1977. Effect of vertical mulch on moisture conservation and yield of sorghum in Vertisols. *Agric. Water Manage.* 1:333-342.

Redinger, G.J., G.S. Campbell, K.E. Saxton, and R.I. Papendick. 1984. Infiltration rate of slot mulches: measurement and numerical simulation. *Soil Sci. Soc. Am. J.* 48:982-986.

Reeve, N.G. and M.E. Sumner. 1972. Amelioration of subsoil acidity in Natal Oxisols by leaching of surface-applied amendments. *Agrochemophysica* 4:1-6.

Sanchez, P.A. and J.G. Salinas. 1981. Low-input technology for managing Oxisols and Ultisols in tropical America. *Adv. Agron.* 34: 280-398.

Saxton, K.E. 1980. Slot mulch tillage for water management and reduced erosion. p. 42-49. In: *Crop production with conservation in the 80's.* Amer. Soc. Agric. Eng., Michigan.

Saxton, K.E. and G.S. Campbell. 1983. Infiltration of slot-mulch tillage. p. 284-291. In: Proc. National Conf. on *Advances in Infiltration.* Amer. Soc. Agri. Eng. publication 11-83, Michigan.

Shainberg, I., M.E. Sumner, W.P. Miller, M.P.W. Farina, M.A. Pavan, and M.V. Fey. 1989. Use of gypsum on soils: A review. *Adv. Soil Sci.* 9:1-11.

Sidorenko, N., N.I. Kartamyghev, and V.A. Poryadin. 1980. The effectiveness of soil slotting. *Zemledelie.* 1:22-25.

Sikora, L.J. and M.E. Radecke. 1987. Future use of sludge entrenchment sites. *Biocycle* 28(8):48-53.

Sommers, L.E. 1977. Chemical composition of sewage sludges and analysis of their potential use as fertilisers. *J. Environ. Qual.* 6 (2): 225-232.

Spain, J.M. and D.L. McCune. 1956. Something new in subsoiling. *Agron. J.* 48:192-193.

Spoor, G. and R.J. Godwin. 1978. An experimental investigation into deep loosening of soils by rigid types. *Agric. Eng. Res.* 23:234-258.

Sumner, M.E. and G.C. Boswell. 1981. Alleviating nutrient stress. p. 99-137. In: G.F. Arkins and H.M. Taylor (eds.) *Modifying the root environment to reduce crop stresses.* Amer. Soc. Agric. Eng., Michigan.

Sumner, M.E. 1970. Aluminium toxicity – A growth limiting factor in some Natal sands. *Proc. S. Afr. Sugar. Technol. Assoc.* 44:197-203.

Sumner, M.E., H. Shahendeh, J. Bouton, and J. Hammel. 1986. Amelioration of an acid soil profile through deep liming and surface application of gypsum. *Soil Sci. Soc. Am. J.* 50:1254-1258.

Swartzendruber, D. 1960. Water movement into soil from idealised vertical mulch channels. *Soil Sci. Soc. Am. Proc.* 24:152-156.

Terzaghi, K. and R. Jelinek. 1954. *Theoretische Bodenmechanik.* Springer-Verlag, Berlin, Gottingen, Heidelberg.

Tiller, K.G. 1989. Heavy metals in soils and their environmental significance. *Adv. Soil Sci.* 9: 113-142.

Unger, P.W. 1975. Relationships between water retention, texture, density and organic matter content of west and south central Texas soils. *Agr. Exp. Stn. Misc. Publ.* No. 1192C: 20.

Vepraskas, M.J. 1988. A method to estimate the probability that subsoiling will increase crop yields. *Soil Sci. Soc. Am. J.* 52: 229-232.

Yushkevich, L.V., V.H. Slesarev, A.R. Makarov, and A.G. Shchitov. 1982. Increasing the effective use of atmospheric moisture in arid agriculture by means of soil slotting. *Soviet Agric. Sci.* 55:42-44.

Amelioration of Subsoil Acidity with Minimum Disturbance

Malcolm E. Sumner

I. Introduction

An enormous volume of literature conveniently summarized by Adams, (1984), Kamprath and Foy (1985); Helyar (1991), and Foy (1992) has been published over the years on the deleterious effects of soil acidity on plant growth and root development. However most of the research has been devoted to the response of crops to the amelioration of acidity in topsoils with relatively little, until recently, focusing on the reduction of acidity in subsoils. Recent review articles (Shainberg et al., 1989; Sumner, 1993) have paid specific attention to the role of gypsum, both mined and byproduct, in improving acid subsoil conditions but few attempts have been made to ascertain the extent to which and the conditions under which lime applied to topsoils is likely to move within the profile. In addition, other potential strategies for ameliorating acid subsoils have not been reviewed.

1-56670-020-5/95/$0.00+.50

Because acid subsoils limit root proliferation over vast areas of the world, this review will briefly outline the nature and extent of acid subsoils throughout the world before proceeding to document the consequences of subsoil acidity in terms of crop productivity. Having formulated the problem, the research work aimed at alleviating the stresses associated with subsoil acidity will be reviewed so that strategies which involve the use of limited or no disturbance of the soil profile in correcting the problem can be evaluated.

No attempt will be made here to cover the literature on subsoil amelioration of artificially constructed profiles in the laboratory or greenhouse except in so far as the results might be illuminating with respect to the mechanisms involved.

II. Nature and Extent of Subsoil Acidity

Basically, subsoils become acid as a result of natural weathering processes or as a result of the intervention of man who, in recent times, has created large acid inputs in the form of nitrogenous fertilizers and acid precipitation (Sumner, 1991). In the former case, the largest areas of such soils are located, predominantly but not exclusively, in humid sub-tropical and tropical regions whereas, in the latter, almost any subsoil can become acid provided sufficient acid inputs are applied. This differentiation will be used in the discussion to follow so that the differences in the chemistry of the soils involved can form the basis of the acidification and amelioration reactions.

A. Humid Subtropical and Tropical Regions

Vast areas in these regions (Southeastern United States, Central and South America, West, Central and Southern Africa, Southeast Asia and the East Indies, Southwest India, Northeastern Australia) contain soils with naturally acid subsoils some of which have become more acid as a result of ammoniacal fertilizer usage.

The reactive fractions of these soils which usually fall in the Oxisol, Ultisol, Alfisol and Andisol Orders of Soil Taxonomy (U.S. Survey Staff, 1975) are basically composed of variable charge surfaces on kaolinite, sesquioxides, amorphous minerals and organic matter. Consequently, acidification of these soils by natural or anthropogenic processes results in a substantial reduction in the cation exchange capacity (CEC) in addition to the normal consequences of acidification, namely, lowered pH and increased levels of soluble Al and Mn. This change in CEC with pH becomes very important in the development of strategies for subsoil amelioration as will be seen later. Because no attempt will be made here to present a complete discussion of acidification processes, interested readers are referred to recent treatises on the subject (Helyar and Porter, 1989; van Breemen, 1991; Bruggenwert et al., 1991; Reuss, 1991).

The decrease in soil solution pH over a period of acidification is the result of a number of reactions as illustrated by (Helyar, 1991):

$$\Delta[H^+] = (pH \mid CEC\ reactions) + (Al\ solid \mid Al^{3+}\ reactions) + (Mn^{4+} \mid Mn^{2+}\ reactions) + (reaction\ of\ CaCO_3\ and\ other\ amendments) + (M^{x+}, Al\ silicates \mid Al\ silicates\ and\ oxides) + (organic\ C\ cycle\ reactions) + (N\ cycle\ reactions) + (acid/alkali\ in\ precipitation\ and\ leaching) \quad (1)$$

As far as the soils in these regions are concerned, the first four terms in the above equation are particularly pertinent in the acidification and amelioration processes. Invariably, subsoils in these regions are naturally acid and the extent of the buffer capacity contributed by variable charge and Al^{3+}, respectively, is illustrated for an Ultisol, Oxic Inceptisol and Oxisol in Figure 1. These curves clearly illustrate the substantial loss in CEC and the increase in exchangeable Al^{3+} as the pH is lowered. Despite the relatively high clay contents of these soils (25-80%), their effective cation exchange capacities (ECEC) are very low (Table 1). Consequently, these subsoils are often very low in basic cations. Acidity in such profiles is often constant or increases with depth. Because the lack of Ca^{2+} and the presence of Al^{3+} in subsoils prevents the elongation of roots (Howard and Adams, 1965; Ritchey et al., 1982; Adams and Moore, 1983; Adams and Hathcock, 1984; Joost and Hoveland, 1986), crop root proliferation in such subsoils is usually very poor. In some soils, Mn^{2+} in addition to Al^{3+} becomes soluble and toxic under acid conditions (Foy, 1992).

B. Humid Temperate and Sub-Humid Subtropical and Tropical Regions

Relatively few of the soils in these regions have subsoils acid enough naturally to present problems as far as deep rooting of crops is concerned. However, some acid subsoils do occur in the Northeastern United States, Eastern and Northern Canada, Northern West Africa, Southern Africa, Southern South America and Northern Europe, examples of which (Humic Eluviated Gleysol, Ustoxic Quartzipsamment) are presented in Table 2. In addition to these naturally acid soils, many areas of neutral to alkaline soils have become acidified by acid rain and the use of nitrogenous fertilizers. In such cases, acidity usually decreases with depth in the profile (Duplex soils). Large areas of anthropogenically acidified soils occur in Southern Australia and Europe, the former being caused by acidification due to legumes and the latter by both acid precipitation and ammoniacal fertilizers.

In these regions, the clay fraction of acid subsoils is usually dominated by layer silicate clay minerals with little or no variable charge. Consequently, the major buffer capacity components of importance in the acid subsoils in these regions would be the Al and Mn reactions in the above equation. Thus on acidification, such subsoils lose little CEC and consequently are much more base rich than the soils in humid subtropical and tropical regions. Thus Ca deficiency is seldom if ever a problem although Al and Mn toxicities frequently occur.

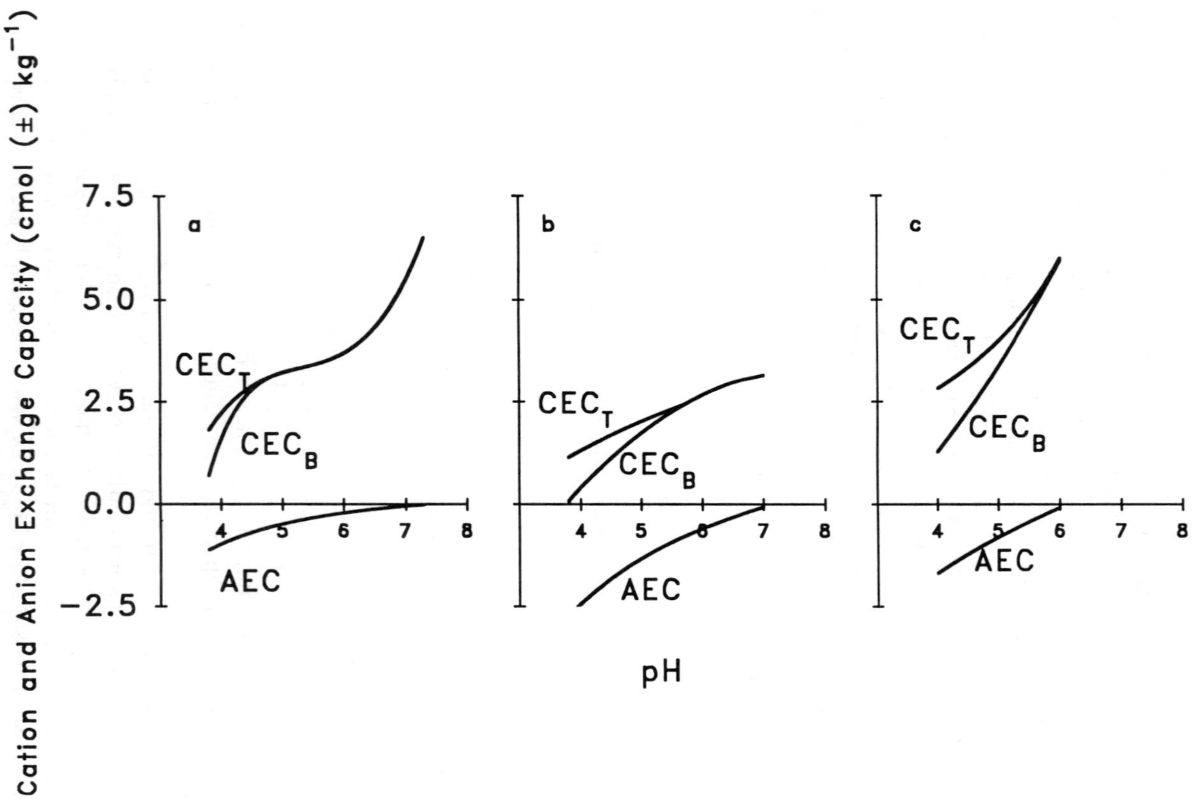

Figure 1. Changes in CEC and AEC with pH for (a) Typic Rhodudult, (b) Oxisol and (c) Oxic Humitropept subsoils (Gillman and Sumner, 1987: Gilllman and Sumpter, 1986; Gillman, 1991). The difference between CEC_T and CEC_B reflects exchangeable Al^{3+}.

Table 1. Profile pH and exchangeable cations in acid soils from humid subtropical and tropical regions

Depth		Exchangeable cations				ECEC
		Ca	Mg	K	Al	
m	pH(KCl)	------------	------------	cmol(+)/kg	------------	------------
Typic Paleudult (Nigeria) (Friesen et al., 1982)						
0-0.15	3.6	0.26	0.09	0.07	1.83	2.86
0.15-0.30	3.7	0.16	0.02	0.02	1.84	2.67
0.30-0.45	3.7	0.15	0.02	0.02	1.94	2.54
0.45-0.60	3.8	0.16	0.02	0.02	2.04	2.73
0.60-0.75	3.7	0.16	0.02	0.02	1.92	2.49
0.75-0.90	3.7	0.14	0.02	0.02	1.94	2.38
Typic Hapludult (Georgia) (Sumner et al., 1986)						
0-0.15	4.4	1.12	0.11	0.08	0.70	2.03
0.15-0.30	3.9	0.70	0.22	0.06	1.48	2.48
0.30-0.45	3.9	0.68	0.33	0.06	1.63	2.73
0.45-0.60	3.9	0.97	0.22	0.05	1.70	2.96
0.60-0.75	3.9	0.65	0.19	0.05	1.96	2.87
0.75-0.90	3.9	0.44	0.10	0.04	2.21	2.81
Typic Acrustox (Brazil) (Ritchey et al., 1982)						
0-0.15	5.0[1]	0.09	0.07	---	0.36	---
0.15-0.30	5.3	0.02	0.04	---	0.14	---
0.30-0.45	5.3	0.02	0.02	---	0.06	---
0.45-0.60	5.4	0.00	0.01	---	0.00	---
0.60-0.75	5.5	0.01	0.01	---	0.00	---
0.75-0.90	5.7	0.01	0.01	---	0.00	---
Oxisol (Queensland, Australia) (Gillman and Sumpter, 1986)						
0.-0.15	4.7[1]	0.60	0.20	0.14	1.10	2.04
0.15-0.30	4.6	0.50	0.10	0.10	1.10	1.80
0.30 0.60	4.6	0.40	0.10	0.07	0.80	1.37
0.60-0.90	4.6	0.50	0.10	0.06	0.60	1.26
Acrudoxic Hapludand (Guatemala) (Sumner et al., 1993)						
0-0.30	4.7[1]	0.55	0.09	0.06	1.20	1.90
0.30-0.60	4.6	0.50	0.08	0.07	0.30	0.95
0.60-0.92	4.8	0.50	0.12	0.03	0.40	1.05
0.92-1.85	4.3	0.50	0.03	0.03	0.50	1.06

[1]pH in water.

Table 2. Profile pH and exchangeable cations in acid soils from temperate and semiarid tropical regions

Depth		Exchangeable cations			
		Ca	Mg	Al	Mn
m	pH	cmol(+)/kg			
Rutland sandy loam (British Columbia, Canada) (Neilsen et al., 1981)					
0-0.10	4.2	3.0	0.8		0.6
0.10-0.20	4.0	1.9	0.6		0.5
0.20-0.30	4.3	2.2	0.9		0.4
0.30-0.40	4.9	4.0	1.1		0.2
0.40-0.50	5.3	5.9	1.2		0.1
0.50-0.60	5.6	5.9	1.1		0.0
Red duplex soil (Victoria, Australia) (Ellington, 1986)					
0-0.10	5.6[1]	1.0	0.3	0.10	0.1
0.10-0.13	5.3	0.8	0.2	0.23	0.1
0.13-0.30	5.6	1.2	0.3	0.12	0.1
0.30-0.75	7.5	5.5	3.3	0.00	0.0
0.75-1.00	7.9	6.9	5.1	0.00	0.0
Yellow duplex soil (Victoria, Australia) (Ellington, 1986)					
0-0.10	5.3[1]	1.0	0.1	0.11	0.17
0.10-0.20	4.9	0.5	0.1	0.72	0.21
0.20-0.30	5.1	1.1	0.2	0.22	0.10
0.30-0.45	6.2	2.5	0.8	0.00	0.00
0.45-0.80	7.3	3.5	2.4	0.00	0.00
0.80-1.10	7.8	4.8	5.1	0.00	0.00
Ustoxic Quartzpsamment (Niger) (Takow et al., 1991)					
0-0.15	4.0	0.2	0.1	0.6	
0.15-0.27	3.9	0.1	0.0	0.7	
0.27-0.44	3.9	0.1	0.0	0.7	
0.44-0.62	3.9	0.0	0.0	0.6	
0.62-0.80	4.0	0.1	0.0	0.6	
0.80-1.03	4.0	0.1	0.0	0.6	
Humic Eluviated Gleysol (Alberta, Canada) (McKenzie and Nyborg, 1984)					
0-0.125	4.4	4.1	0.8	7.8	
0.125-0.25	4.2	4.5	1.1	11.6	
0.25-0.50	4.0	5.3	1.3	10.3	

[1]pH in water.

III. Consequences of Subsoil Acidity

Because acid subsoils containing toxic levels of Al or deficient amounts of Ca restrict root proliferation, the major consequence of subsoil acidity manifests itself as drought stress when the ameliorated topsoil dries out reflecting the lack of roots in zones of the soil where water is available (Bradford and Blanchar, 1977; Doss et al., 1979; Hammel et al., 1985; Sumner, 1990; Sumner, 1993). In addition, acidity can reduce the uptake of nutrients from the subsoil where reservoirs might be present (Sumner et al., 1991).

IV. Amelioration with Some Disturbance

Much work (Lathwell and Peech, 1964; Bradford and Blanchar, 1977; Perez-Escolar and Lugo-Lopez, 1978; McKenzie and Nyborg, 1984; Hammel et al., 1985; Sumner et al., 1986) has been conducted in which lime has been mechanically mixed into the soil to great depths (0.5-1 m) to neutralize subsoil acidity. In all cases, such treatments have proven to be highly effective in promoting deep rooting resulting in higher yields. Because such approaches are impractical and highly uneconomic, they will not be discussed here. Only techniques which are likely to prove to be feasible in practical agriculture will be reviewed. The discussion to follow will, therefore, focus on techniques in which lime is mixed with only a limited volume of soil or where ameliorated topsoil is buried in vertical bands in the subsoil.

Doss et al. (1979) successfully used a chisel plow (0.45 m) in combination with a rotary tiller (0.23 m) to incorporate lime into a Rhodic Paleudult and a Typic Paleudult. As a result of the neutralization of acidity, rooting depth, soil water extraction and yields of cotton and corn were increased by these treatments. Similar results were obtained by Gonzalez-Erico et al. (1979) on a Typic Haplustox by incorporating lime to a depth of 0.3 m. Coventry et al. (1987) found that deep ripping (0.4 m) in combination with topsoil incorporated lime was highly effective in increasing wheat yields on a duplex soil in Australia acidified over the years by legume N.

Anderson and Hendrick (1983) developed a subsoil lime injector consisting of a slitter blade attached to a chisel which successfully applied suspended lime along the sides of the vertical slit to a depth of 0.36 m. However, no results of improved rooting as a result of this treatment were presented. de Roo (1956) produced a machine for deep placement of lime which resulted in improved rooting to approximately 0.20 m below furrow bottom.

Lutz and Jones (1971) using a moldboard plow, placed lime (22.4 Mg/ha) and fertilizer (244 kg P/ha and 930 kg K/ha) in different combinations on the plowsole at a depth of 0.30 m in a Davidson clay loam (Rhodic Paleudult) in 1969. Two years later, there was no evidence of lime moving from the point of placement. They found no yield response to plowsole lime in soybeans which, however, did respond to the deep placed nutrients in the third year. In 1978, the same plots were planted to alfalfa by Rechcigl et al. (1985a,d). Plowsole application of lime resulted in increased alfalfa root penetration from 0.30 to 0.60 m and yield increases from 7.1 to 11.7 Mg/ha on average over a 5 year period with corresponding increases in pH and decreases in exchangeable Al below the zone of placement. Rechcigl et al. (1985b,c) attributed the differences in behavior of the two crops (soybeans and alfalfa) to plowsole lime and nutrients to differential rooting patterns or Al tolerance or both. The data of Rechcigl et al. (1985b,c) indicate that some movement of lime below the plowsole took place over the 13 year period since application. They concluded that the residual effect of plowsole placement of lime was a viable means of encouraging deep rooting on soils with acid subsoils. Pohlman (1946), on a strongly acid Gilpin silt loam (Typic Hapludult), also demonstrated lime movement 9 years after application. When the topsoil was limed to pH 7, lime moved 0.20 m beyond point of placement whereas at lower lime rates (pH 6) in the topsoil, movement was much less (0.10 m). Depth of lime incorporation strongly affected alfalfa yields in this experiment.

Gonzalez-Erico et al. (1979) who compared the incorporation of lime to 0.15 and 0.30 m in a Typic Haplustox in Brazil, obtained highly significant yield increases to the deeper mixing due to reductions in toxic Al which promoted more prolific rooting and water extraction from the subsoil. Unfortunately, the experiment ran for only 3 years and the data presented indicated little or no downward movement of the lime beyond the point of placement.

Saayman and van Huyssteen (1981) developed a lime applicator to blow lime into the soil behind various tillage implements. They found that lime applied on the surface before deep plowing gave surprisingly good results in comparison to lime blown in with the applicator. Double plowing was an efficient method to ensure proper mixing of the lime.

Farina and Channon (1988a) compared various mechanical procedures for ameliorating subsoil acidity on a Plinthic Paleudult having an acid saturation (Al+H) increasing from 59 to 75% and a Ca saturation decreasing from 19 to 10% with depth. pH in KCl was almost constant at 4.0 over the top 0.90 m of depth. Among the most successful treatments were a deep limer comprising a subsoiler equipped with a lime applicator and an orchard fan to force lime into a 0.07 m wide vertical slot to a depth of 0.75 m, and a modified subsoiler which facilitates the incorporation of previously limed topsoil in 0.065 m wide vertical bands to a depth of 0.7 m below the surface. Both these implements at 0.9 m centers had a diesel fuel requirement of approximately 40 L/ha using a 93 kW four-wheel drive tractor. Their results are presented in Table 3 in which these treatments are compared with other more expensive less practical mechanical operations to incorporate lime to depth. Clearly, these subsoiler treatments are highly effective and potentially economic being currently used widely by farmers in Natal, South Africa. By comparison, surface incorporation of gypsum was even more effective than any mechanical treatment to incorporate lime. This will be discussed later in Section V.D. Using a similar strategy, Coventry (1991)

Table 3. Comparison of various deep tillage and lime treatments on the cumulative corn yield increase over conventional tillage for the period 1983/84 to 1990/91 on a Plinthic Paleudult in Natal, South Africa

Treatment	Corn Yield (kg ha^{-1})
Limed topsoil + conventional tillage	0
Limed topsoil + deep plow (Nardi)	6816
Limed topsoil + modified subsoiler (90 cm)	4216
Limed topsoil + deep limer (90 cm) with extra 14 Mg lime ha^{-1}	4752
Limed topsoil + lime (10 Mg^{-1}) in furrow with Wye double digger	7096
Limed topsoil + surface gypsum (10 Mg ha^{-1})	10272

(From Farina and Channon, 1988. Updated figures kindly supplied by Dr. Farina.)

obtained yield responses in wheat to the injection of lime slurries (5 Mg/ha) at 0.35 m using a modified paraplow on acid duplex soils in Australia.

Kirchhof et al. (1991) used a slotting machine, originally developed to overcome infiltration problems in sodic soils (Jayawardane and Blackwell, 1985), to incorporate lime mixed with soil in 0.15 m wide slots 0.4 m deep in a commercial vineyard. They observed more than a tenfold increase in root length density in the slotted limed soil over the control areas and while a 72% increase in yield in the vineyard was observed the contribution of slotting and other management practices could not be separated. However as the vineyard has been observed to be less subject to short term water deficits subsequent to treatment, a large part of the yield increase is likely to be due to the improved soil conditions in the slots.

Ellington et al. (1992) obtained substantial yield increases in wheat after injecting a lime slurry at 10-35 cm below the surface of a solodic soil in Australia. This treatment reduced exchangeable Al and increased pH. Gypsum in combination with lime was better than lime alone and their data suggested that gypsum might promote the downward movement of lime.

Previous work by Pearson et al. (1973) in root boxes in the greenhouse demonstrated that even when lime was poorly mixed with very acid subsoil (Ultisol) rooting depth was increased at least twofold.

The conclusion can, therefore, be drawn from the above studies that, the deeper lime is incorporated into an acid subsoil, the better is the root proliferation likely to be in the ameliorated zone which usually translates into increased yields. However, the practical efficacy of such treatments is determined by economic considerations which depend on soil type, crop, prices, tillage costs, etc. and therefore a general rule cannot be formulated for all situations.

V. Amelioration without Disturbance

A. Surface Applied or Topsoil Incorporated Lime without Acid Inputs

There are conflicting reports on the efficacy of surface applied lime in neutralizing subsoil acidity. A summary of some of these observations is presented in Table 4 from which the diversity of findings can readily be seen. In some cases, no movement even after considerable periods of time had elapsed since liming was observed whereas in others movement appears to be quite rapid. There appears to be no unifying thread through these observations on which predictions can be based. The problem with many of these observations is that the conditions under which the work was carried out was often not described in sufficient detail for meaningful interpretations to be made. In addition, the possibility exists of contaminating subsoil with treated topsoil in the process of sampling particularly when using bucket type augers. Soil heterogeneity is also likely to be a major contributing factor. In addition, in many cases, the movement of Ca has been used as the index of lime movement which is not necessarily true. On the contrary it is the movement of alkalinity which is important in increasing subsoil pH. Farina (1993) has demonstrated appreciable movement of Ca and Mg from lime applied in the topsoil but with no measurable reduction in Al+H in the subsoil. These apparent contradictions in the way in which lime moves in soil are nothing new. This state of confusion has persisted for the past 70 years as can be judged by the words of Stewart and Wyatt (1919) in describing the results of extensive liming studies: "Limestone applied to the surface slowly penetrates into the subsurface... Applications of limestone to the surface soil seem to have no effect on the acidity of the subsoil." These statements adequately sum up the data presented in Table 4.

Quite clearly, sufficient is now known about the chemistry of acidity in permanent and variable charge soils (Uehara and Gillman, 1981; Adams, 1981; Thomas and Hargrove, 1984; Adams, 1984; Sposito, 1989; Gillman, 1991; Helyar, 1991; Ulrich and Sumner, 1991) to attempt to formulate certain rules which might be useful in predicting the likely behavior of lime within the soil profile. In order for lime applied on the surface or incorporated into the topsoil to have an influence on subsoil acidity, alkalinity usually in the form of HCO_3^- or OH^- must be transported by mass flow from surface horizons in a downward direction. Of course, acidity in the form of H^+ or acidic cations such as Al^{3+}, Mn^{2+} and to a very limited extent Fe^{2+} can also be transported in a similar manner. In the pH range 5.2 to 5.6, mass flow balance exists between alkalinity and acidity because of the insolubility of Al, Fe and Mn and the equality of HCO_3^- and H^+ concentrations at these pH values (Ritchie, 1989; Helyar, 1991). Therefore for significant alkalinity to be mobile, the pH in the topsoil must be well above 5.6 where the concentrations of HCO_3^-, OH^- and CO_3^{2-} begin to increase logarithmically.

In acid subsoils which are dominated by permanent charge minerals (soils of humid temperate and semiarid tropical and subtropical regions), the only further constraint to the movement of alkalinity is the requirement that the acidity present in the subsoil be neutralized as the alkaline front moves downward. Given sufficient excess alkalinity in the topsoil, the rate of downward movement should be proportional to the acid content of the subsoil which is usually present in the form of Al^{3+} and Mn^{2+}. The following reaction describes the process taking place in the subsoil:

$$2\ \text{Exch-}Al^{3+} + 3\ Ca(HCO_3)_2 \Rightarrow 3\ \text{Exch-}Ca^{2+} + 2\ Al(OH)_3 + 6\ CO_2 \quad (2)$$

Thus Ca^{2+} stoichiometrically replaces Al^{3+} on the existing permanent charge sites whose magnitude does not change under the influence of the increased alkalinity.

On the other hand in variable charge subsoils (soils of the humid tropics and subtropics and andisols), the advancing alkaline front would be further retarded by the additional requirement that a portion of the available alkalinity must be used to develop pH dependent charge which can be quite sizeable in some highly weathered soils. The processes taking place in this situation are illustrated in the following reactions:

$$2\ \text{Exch-}Al^{3+} + 3\ Ca(HCO_3)_2 \Rightarrow 3\ \text{Exch-}Ca^{2+} + 2\ Al(OH)_3 + 6\ CO_2 \quad (3)$$

$$\text{Solid-OHH}^{+}Cl^{-} + \tfrac{1}{2}\ Ca(HCO_3)_2 \Rightarrow \text{Solid-OH} + H_2O + CO_2 + \tfrac{1}{2}\ CaCl_2 \quad (4)$$

$$\text{Solid-OH} + \tfrac{1}{2}\ Ca(HCO_3)_2 \Rightarrow \text{Solid-O}^{-}\tfrac{1}{2}Ca^{2+} + H_2O + CO_2 \quad (5)$$

The magnitude of Reactions 4 and 5 will depend on the nature and extent of the variable charge component in the subsoil. Note that when subsoils contain positively charged sites as in Reaction 4, salt ($CaCl_2$) which would be free to move downward rapidly, is generated. This explains the results of some workers (Messick et al., 1984) [Typic Hapludult], (Friesen et al., 1982) [Typic Paleudult], (Juo and Ballaux, 1977) [Oxic Paleudult] who reported rapid downward movement of Ca^{2+} in such soils with little or no change in pH.

Thus in variable charge subsoils at a given acid content, slower movement of alkalinity down the profile than that observed in permanent charge subsoils, would be expected. Even in the latter category of soils, movement has been calculated to be very slow. Porter and Helyar (1992), on the basis of extensive acidification rate studies on Australian soils, have estimated that it is likely to take between 20 and 100 years for the pH to be raised by 0.5 in the 0.1 m layer underlying a limed topsoil. Of course, raising the topsoil pH to high levels would accelerate this rate but the growth of crop plants might suffer particularly on variable charge soils (Farina et al., 1980; White and Robson, 1989).

In certain cases, significant downward movement has been reported even when the pH of the topsoil was below 6 suggesting that a mechanism other than the transfer of HCO_3^- in the mass flow might be responsible. Helyar (1991) proposed that, in such a case, mass flow in both surface and subsurface layers is sufficiently acid before liming to contain significant excesses of Al^{3+}, Mn^{2+} and

Table 4. Examples of cases of differential movement of lime within the soil profile

Location	Soil type	Lime rate (Mg ha^{-1})	Years after liming	Movement (m)	Source
Queensland, Aust.	Oxisol	2	<1	0	Gillman, 1991
Virginia	Rhodic Paleudult	22.4 placed at 0.3 m	9	0	Recheigl et al., 1991
Malaysia.	Typic Paleudult	2	0.3	0	Shamshuddin et al., 1991
Virginia	Typic Hapludult	22.4 placed at 0.3 m	3	0	Rechcigl et al., 1985
Virginia	Typic Paleudult	13.4	4	0	Messick et al., 1984
Virginia	Aquic Fragiudult	6.7	4	0	Rechcigl et al., 1985
British Columbia	Rutland sandy loam	6	4	0	Neilsen et al., 1981
Hawaii	Typic Hydrandept	10 and 19	5	1.0	Fox et al., 1991
São Paulo, Brazil	Typic Haplorthox	5.5	0.3	0.2-0.6	Caires and Rosolem, 1991
NSW, Australia	Granite derived	3.6	10	0.2	Cumming, 1991
NSW, Australia	Basalt derived	5	10	0.2	Cumming, 1991
Jamaica	Cunupia sandy loam	15	4	0.3	Weir, 1975
Virginia	Typic Hapludult	13.4	4	0.3	Messick et al., 1984
Virginia	Typic Hapludult	13.4	4	0.2	Messick et al., 1984
Victoria, Australia	Yellow podzolic	11.25	40	0.5	Ridley et al., 1990
Zambia	Oxic Paleustult	8	6.5	0.5	McKenzie et al., 1988
Connecticut	Typic Dystrochrept	16	20	0.5	Brown et al., 1956
Tasmania, Aust.	Kraznozem	15	1.8	0.2	Rowe, 1982
Nigeria	Typic Paleudult	4	5	0.8	Friesen et al., 1982
NSW, Australia	Yellow duplex	5.6	12	0.3	Bromfield et al., 1987

Table 4. continued

Location	Soil type	Lime rate (Mg ha^{-1})	Years after liming	Movement (m)	Source
Hawaii	Typic Hydrudand	17	5	1.0	Mahilum et al. (1970)
South Africa	Kipipiri	6	6	0.5	Meyer et al. (1991)
Georgia	Typic Kanhaplucult	pH[1] 6.5	6	0.75	Adams et al. (1967)
New Hampshire	Dystric Eutrochrept	9	6	0.2	Koch and Estes (1986)
Brazil	Typic Hapludox	?	3.5	0.5	Goedert (1983)
Connecticut	Typic Dystrochrept	10	18	0.9	Brown and Munsell (1938)
NSW, Australia	Solodized solonetz	8	5	0.1	Conyers and Scott (1989)
NSW, Australia	Red earth	5.5	13	0.2	Horsnell (1985)
South Africa	Plinthic Paleudult	25	10	<0.1	Farina (1993)

[1] Topsoil maintained at constant pH.

H^+. Liming the surface to pH 5.5 would eliminate the transfer of acid from the topsoil to subsoil but not the acid output from the subsoil. Thus the pH in the subsoil would rise slightly with time.

Furthermore, the mixing action of soil fauna (Springett, 1983; Lobry de Bruyn and Conacher, 1990), has been shown to result in the transfer of alkalinity within soil profiles. Certain earthworms (*Allolobophora caliginosa* and *Lumbricus rubellus*) tend to mix lime laterally whereas *Allolobophora longa* mixes it vertically (Figure 2).

Seldom, if ever, would lime be applied to a topsoil with the objective of improving the subsoil without any acid inputs in the form of nitrogenous fertilizers or legumes. Without acid inputs, transfer of alkalinity to the subsoil will be slow. Consequently, the potential for lime movement into the subsoil must be considered in the case of acid inputs in order to effectively estimate its contribution.

B. Surface Applied or Topsoil Incorporated Lime with Acid Inputs

Many examples have been presented in the literature to demonstrate that alkalinity can be transferred to subsoils by the addition of lime and acidifying fertilizers to topsoils (Percival et al., 1955; Pearson et al., 1962; Abruna et al., 1964; Weir, 1975; Adams, 1981). This can be brought about rapidly by acid inputs from ammoniacal nitrogen fertilizers to well limed topsoils and the leaching of the resulting $Ca(NO_3)_2$ and other salts formed to the subsoil where the more rapid rate of assimilation of NO_3^- than Ca^{2+} by roots causes the subsoil pH to increase. This strategy is well illustrated by the data of Adams et al. (1967) in Figure 3. With no or insufficient lime, the high rate (900 kg/ha/yr) of N as NH_4NO_3 markedly acidified the topsoil but in the presence of adequate lime, the pH of the entire profile (0.8 m) was raised above the original level. Similar rapid effects can be obtained by the use of high rates of $Ca(NO_3)_2$ on forages as demonstrated by Adams and Pearson (1969). However with lower rates of $Ca(NO_3)_2$ (224 kg N/ha), Perez-Escolar and Lugo-Lopez (1978) obtained no responses in four successive crops of corn grown on a Typic Tropohumult. Leaching soils treated with $Ca(NO_3)_2$ in the absence of plants is unlikely to be very effective as demonstrated by Kotze and Deist (1975) and Pleysier and Juo (1981) who showed that exchangeable Al and Ca were only slightly decreased and increased respectively in the subsoil. With certain acid intolerant crops, roots may not be present in sufficient quantities in the subsoil to make this strategy a viable option in the short term. However, the use of acid tolerant crops such as tropical grasses would tend to accelerate the transfer of alkalinity from topsoil to subsoil. Pleysier and Juo (1981) found that the NO_3^- leached from the soil was accompanied by Al^{3+}.

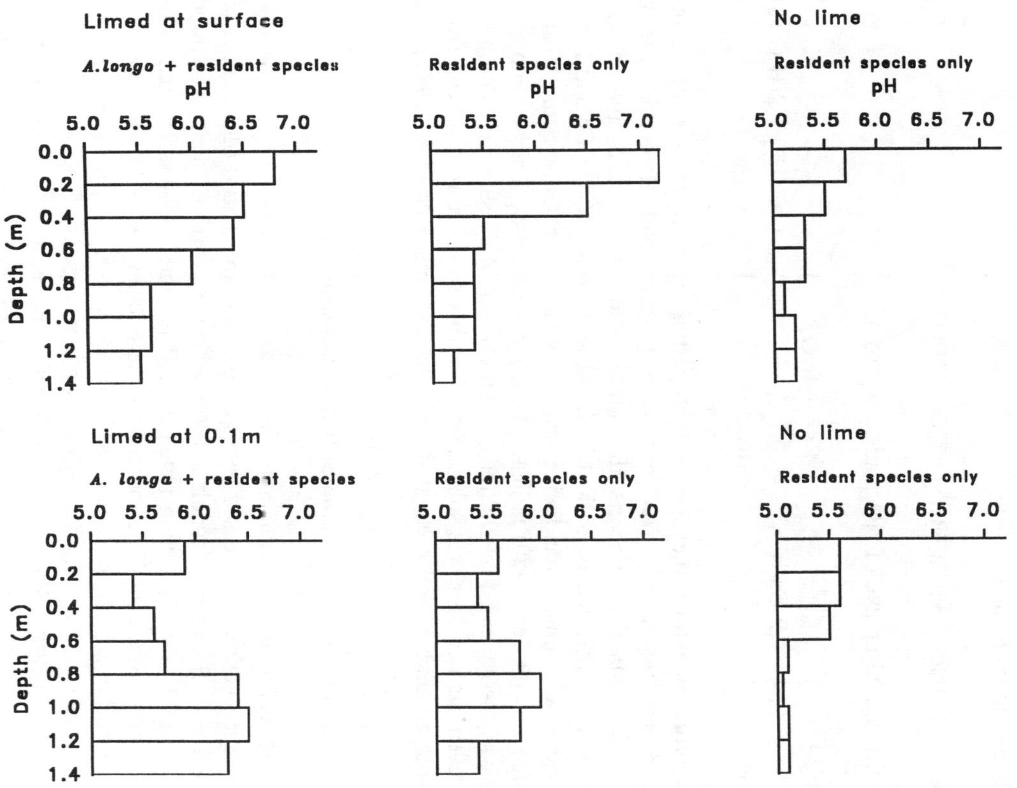

Figure 2. Effect of introduction of *Allolobophora longa* on pH profiles of a Fragiaqualf where lime (5 Mg ha^{-1}) was either surface applied or incorporated to a depth of 0.1 m. (From Springett, 1983.)

When this strategy is adopted, the following reactions illustrate the processes taking place:

(1) Nitrification

$$NH_4X + 2\ O_2 \Rightarrow HNO_3 + HX + H_2O \qquad (6)$$

where X can be NO_3, $\frac{1}{2}SO_4$ or Cl

(2) Reaction with lime

$$CaCO_3 + 2\ HNO_3 \Rightarrow Ca(NO_3)_2 + CO_2 + H_2O \qquad (7)$$

(3) Differential uptake of NO_3 and Ca by roots

$$\left.\begin{matrix} \text{Root} \rightarrow 4OH^- \\ \text{Root} \rightarrow 2H^+ \end{matrix}\right\} + 2\ Ca(NO_3)_2 \Rightarrow \left.\begin{matrix} \text{Root} \leftarrow 4NO_3^- \\ \text{Root} \leftarrow Ca^{2+} \end{matrix}\right\} + Ca(OH)_2 + 2\ H_2O \qquad (8)$$

Other work in which the chemical properties of soils from paired sites (cultivated and woodland) were compared (Adams and Hathcock, 1984), although circumstantial, supports this mechanism. In all cases the subsoil horizons in the cultivated fields had high pH and exchangeable Ca and lower exchangeable Al contents than their paired woodland sites. Other data (Adams, 1981; Manson and Fey, 1989) indicates that, if Cl^- and NO_3^- leached into acid subsoils was accompanied by Na^+ rather than Ca^{2+} or Mg^{2+}, subsoil solution pH may be even higher. This was substantiated by better maize root growth from topsoil into an acid subsoil pretreated with NaCl (Manson and Fey, 1989).

C. Movement of Organic Compounds into Subsoils

Lund and Doss (1980) found that application of 135 Mg of animal manure/ha/yr for three years markedly increased levels of exchangeable K and Mg, and pH in the subsoil of a Plinthic Paleudult and the effect was still evident after three years. Hern et al. (1982) demonstrated that, if an acid soil was leached with EDTA in the presence of lime, nearly all the exchangeable Al was removed whereas topsoil treatments of lime alone had little effect. Their data suggest that organic ligands can have a marked effect on the movement of metal ions in soil systems. Subsequently, Wright et al. (1985) showed on a Typic Hapludult that, when the topsoil was treated with cow manure (44.8 Mg/ha) or EDTA (5.1 Mg/ha) in the presence of dolomitic lime (8.4 Mg/ha) and leached with water, subsoil pH and exchangeable Al increased and decreased, respectively, whereas lime alone had no effect. However, rooting of an acid sensitive barley variety was best in the treatments receiving manure with or without lime. Their results suggest that the response was due to the complexation of toxic Al^{3+} by soluble

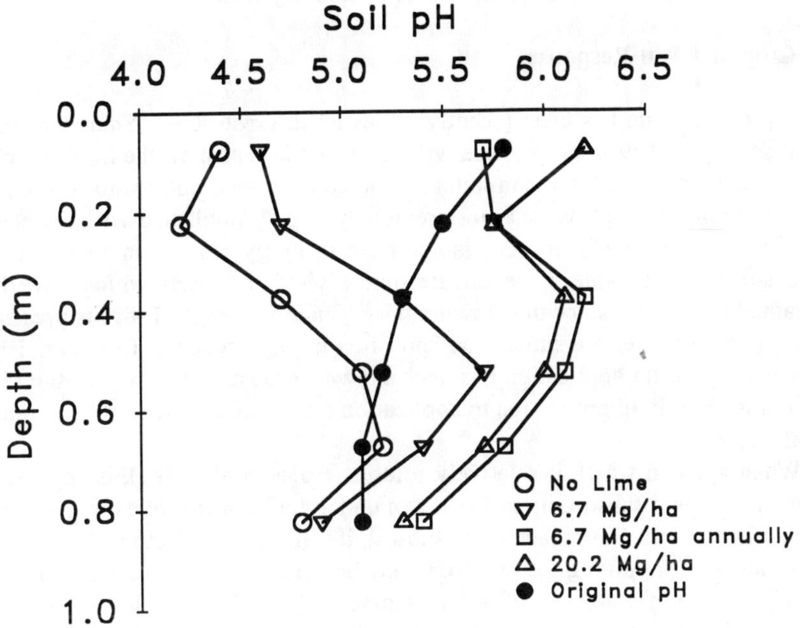

Figure 3. Effect of rates of calcitic limestone on soil profile acidity after 4 years under a Coastal bermudagrass (*Cynodon dactylon* [L.] Pers.) sod fertilized with NH_4NO_3 at an annual N rate of 900 kg/ha. Single lime applications were made at the beginning of the experiment except for the treatment labeled "6.7 Mg/ha annually." (From Adams et al., 1967.)

organic components. In solution culture experiments with cotton, Hue et al. (1986) demonstrated that short chain carboxylic acids can detoxify Al^{3+} to varying extents depending on the relative position of the OH/COOH groups on their main C chain favoring the formation of 5- and 6-bond ring structures with Al. When soil solutions from acid eluviated horizons of various Ultisols were investigated, several organic acids were identified with higher concentrations being obtained in material from forested rather than cultivated sites. After partitioning total Al concentrations in the soil solutions into organic complexed and monomeric Al, cotton root growth was significantly correlated with monomeric Al and not with total Al in solution. These observations indeed support the view that soluble organic compounds in soils are mobile and capable of detoxifying acid subsoils. In many areas of the world, large quantities of organic wastes (chicken litter, sewage sludge, hog and cow manure) are applied to land which often is acid in the subsoil. The extent to which these materials might be beneficial in improving subsoil conditions has not been adequately investigated. This is a potentially fruitful avenue for further research.

D. Surface Applied or Topsoil Incorporated Gypsum

1. Crop and Soil Responses

Because this topic has been recently reviewed in great detail (Shainberg et al., 1989; Sumner, 1993), no purpose will be served in repeating the information so that readers interested in the minutia should consult these publications. Rather in the discussion to follow, data not previously widely published will be selected to illustrate the benefits to be obtained from using gypsum as an ameliorant for acid subsoils. In Table 5, an update of the yield responses which have been obtained to gypsum since the last summary (Shainberg et al., 1989) is presented. The yield responses are substantial and often highly economic (Sumner, 1990). However, gypsum application has been shown to substantially reduce the yields of lupins roughly in proportion to application rate (McLay et al., 1992; Ellington et al., 1992).

When gypsum which is relatively soluble (Bolan et al., 1991), is applied to a topsoil with acid subsoil, it leaches down under the influence of excess moisture. When the gypsum front reaches the subsoil, it fulfills one or both of two possible roles, namely, supplying Ca^{2+} where it may be deficient and reducing the toxicity of Al^{3+} in one of a number of ways (Sumner, 1993). As a result of this subsoil amelioration, roots are able to proliferate in previously unavailable zones enabling them to more efficiently extract water. This is graphically illustrated in Figure 4 for alfalfa roots in a Typic Halpudult profile in Georgia, eight years

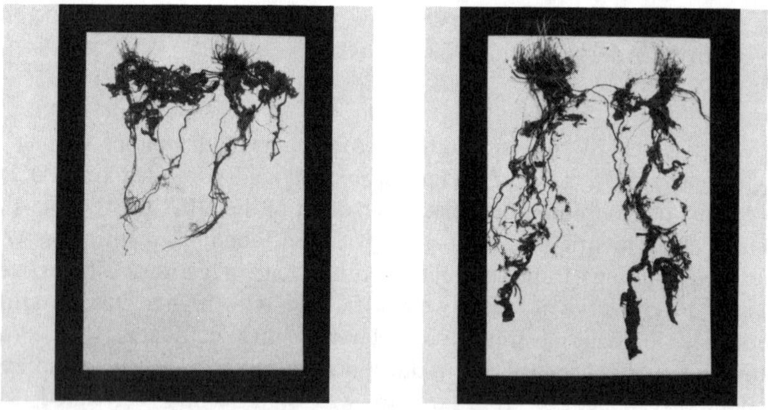

(a) **(b)**

Figure 4. Comparison of rooting patterns in an Appling soil (Typic Hapludult) treated with 10 Mg/ha of gypsum 8 years previously. (a) Control, (b) Gypsum. (From Sumner, 1990.)

Table 5. Examples of yield responses in various crops to surface applications of topsoil incorporated gypsum

Crop	Location	Soil type	Gypsum rate ($Mg\ ha^{-1}$)	Yield response %	Source
Corn	South Africa	Plinthic Paleudult	10	19	Farina and Channon (1988)
Coffee	Brazil	Latosol Roxo (Rhodic Hapludox)	2.6	41	Guimaraes (1992)
Sugarcane	Brazil	Dark Red Latosol (Typic Hapludox)	6	7	Vitti et al. (1992)
Sugarcane	Brazil	Arenoso Distrofico	2	16	Vitti et al. (1992)
Sugarcane	Brazil	Red Yellow Latosol (Typic Hapludox)	6	8	Vitti et al. (1992)
Corn	Brazil	- -	8	9	Quaggio (1992)
Corn	Brazil	Red Yellow Latosol (Xanthic Hapludox)	6	76	Souza et al. (1992)
Corn	Brazil	Red Latosol (Rhodic Hapludox)	4	47	Souza et al. (1992)
Wheat	Brazil	Red Yellow Latosol (Xanthic Hapludox)	6	72	Souza et al. (1992)
Leucena	Brazil	Red Yellow Latosol (Xanthic Hapludox)	6	81	Souza et al. (1992)
Sugarcane	Brazil	Acid sand (Udoxic Quartzipsamment)	5.6	17	Dematte (1992)
Sugarcane	Brazil	Acid sand (Udoxic Quartzipsamment)	6	7	Dematte (1992)
Corn	Brazil	Dark Red Latosol (Typic Hapludox)	6	19	Ernani et al. (1992)
Barley	Brazil	Dark Red Latosol (Typic Hapludox)	6	11	Ernani et al. (1992)
Beans	Brazil	Latossolo Bruno Distrofico (Typic Hapludox)	6	13	Ernani et al. (1992)
Oats	Brazil	Dard Red Podzolic (Rhodic Kandiudult)	1	32	Ernani et al. (1992)
Corn	Brazil	Dark Red Latosol (Typic Hapludox)	6	82	Malavolta (1992)
Corn	Brazil	Dark Red Latosol (Rhodic Kandiudult)	6	77	Malavolta (1992)
Coffee	Brazil	Cerrado (Oxisol)	2.6	59	Malavolta (1992)
Alfalfa	Georgia	Typic Kanhapludult	10	18	Riley (1991)
Sugarcane	South Africa	Humic Hapludox	5	13	Meyer et al. (1991)
Wheat	Australia	Yellow Sandplain	9	55	McLay et al. (1992)
Pasture	Australia	- -	2.5	28	Peoples et al. (1992)
Soybean	Kentucky	Typic Hapludult	3.5	40	Marsh and Grove (1992)

Table 6. Effect of surface incorporated gypsum on the yield of corn and soybeans grown on a Red Yellow Latosol (Xanthic Hapludox) in the Campo Cerrado of Brazil during the wet and dry seasons

	Yield					
Gypsum applied	Corn in dry season (1982) Days without irrigation			Soybeans in wet season	Corn in dry season	Soybeans in wet season
	0	21	33	(1982/83)	(1983)	(1983/84)
	Mg ha^{-1}					
0	3.50	1.71	0.27	2.82	2.83	2.56
2	--	2.77	0.70	2.83	3.22	2.69
4	--	2.82	0.60	2.89	4.05	2.77
6	4.17	3.11	0.72	2.75	4.94	2.65

(From Carvalho et al., 1986.)

after treatment. The improved water availability translates, in turn, into higher yields. Obviously, such yield increases are likely to be larger during droughty than well watered seasons as illustrated in the data of Carvalho et al. (1986) in Table 6. The beneficial effect of gypsum in the two dry seasons (1982 and 1983) on the yield of corn is clearly evident even when the crop was irrigated. However in the wet seasons (1982/83 and 1983/4) during which there were no moisture deficits, soybeans showed no response. The better distribution of corn roots in the lower part of the profile in the 1983 dry season is apparent in Figure 5 which is corroborated by the volumetric water content remaining in the soil during a dry period without irrigation in Table 7. The moisture content in the upper 0.2 m of soil in all treatments is at or below the wilting point. In the control treatment, the soil below 0.4 m is quite wet which contrasts with the much drier conditions down to 1.2 m in the 6 Mg gypsum/ha treatment. These data reflect the deeper rooting pattern in the gypsum treated soil. Farina (1993) suggested that the differences in response in the data of Carvalho et al. (1986) might not be due only to seasonal moisture differences but also to the remarkable ability of soybeans well supplied with Mo to penetrate acid subsoils. His data to support this view are presented in Table 8 where the performance of maize and soybeans in a well watered (1990/91) and a severe drought (1991/92) season are compared in a split plot experiment on a Plinthic Paleudult. In the season adequately supplied with water (1990/91), maize responded to gypsum wheareas soybeans did not. However, in the severe drought season (1991/92), maize failed totally while the soybeans yielded only slightly less than in 1990/91 but did respond to gypsum when not treated with Mo. Plots receiving Mo showed no response to gypsum (data not presented). In a long-term experiment (Farina and Channon, 1988) in which mechanical incorporation of lime was compared with surface applied gypsum, the latter, after a period of years, became the highest yielding treatment in the experiment (Table 3). The gypsum treatment had more

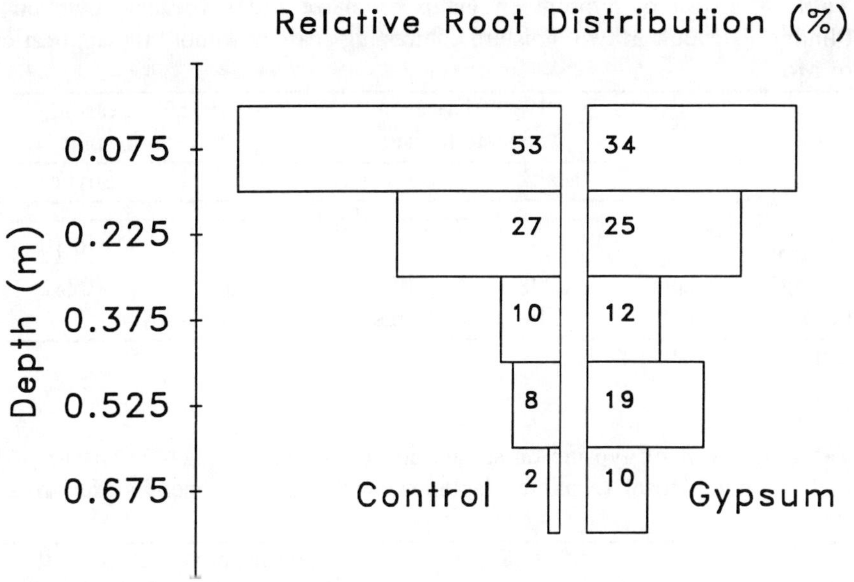

Figure 5. Effect of gypsum (6 Mg/ha) on the relative distribution of corn roots with depth in a Red Yellow Latosol (Xanthic Hapludox) 90 days after emergence in the 1983 dry season. (From Souza and Ritchey, 1986.)

Table 7. Effect of gypsum on the volumetric water content of a Red Yellow Latosol (Xanthic Hapludox) with depth under a corn crop 75 days after emergence at the end of a 25 day period without irrigation in the dry season of 1983

Depth	Volumetric water content of soil at		Volumetric water content remaining in soil at gypsum rate indicated (Mg ha^{-1})			
m	0.1 bar	15 bar	0	2	4	6
	m^3/m^3					
0-0.1	0.293	0.181	0.083	0.094	0.142	0.142
0.1-0.2	0.293	0.181	0.189	0.186	0.189	0.189
0.2-0.3	0.293	0.181	0.190	0.190	0.190	0.202
0.3-0.4	0.269	0.180	0.185	0.196	0.196	0.196
0.4-0.6	0.269	0.180	0.227	0.205	0.200	0.211
0.6-1.0	0.269	0.181	0.269	0.237	0.217	0.208
1.0-1.2	0.269	0.181	0.269	0.269	0.220	0.220

(From Souza and Ritchey, 1986.)

Table 8. Effect of gypsum on yields of maize and soybeans grown on a Plinthic Paleudult in two moisture contrasting seasons without the application of Mo

	1990/91 Season Adequate moisture		1991/92 Season Severe drought	
	Maize	Soybeans	Maize	Soybeans
	---------------------------kg ha^{-1}	---------------------------	---------------------------	---------------------------
Control	5473	2956	0	2012
10 Mg $^{-1}$ Gypsum	6208	3053	0	2282
$LSD_{0.05}$	243	n.s.		167

(From Farina, 1993.)

Table 9. Effect of gypsum on sulfate adsorbed and net negative charge of a Red Yellow Latosol (Xanthic Hapludox) with depth 11 months after treatment

Depth m	Sulfate (SO_4^{2-}) absorbed	Net negative charge Gypsum rate (Mg ha^{-1}) 0	6
	---------------------cmmol$_c$ kg^{-1}	---------------------	---------------------
0-0.15	0.12	2.87	3.65
0.15-0.30	1.00	1.11	1.42
0.30-0.45	1.30	1.04	1.15
0.45-0.60	1.40	0.74	1.13
0.60-0.75	1.30	0.83	1.13
0.75-0.90	0.62	0.58	0.91
0.90-1.05	0.25	0.40	0.65
1.05-1.20	0.05	0.31	0.41

(From Souza and Ritchey, 1986.)

than doubled root length in the 0.6-0.9 m layer over the control in 1992 (Farina, 1993). These results indicate that, provided acidity in the topsoil is neutralized with lime, gypsum is very effective in offsetting the deleterious effects of subsoil acidity. Comparison of the response to gypsum to that of the various mechanical procedures of incorporating lime (Table 3) indicates that, in the long term, gypsum is likely to be more effective in neutralizing the deleterious effects of subsoil acidity.

Because the Red Yellow Latosol (Xanthic Hapludox) used in the above experiment (Carvalho et al., (1986) has a large variable charge component, gypsum increases the net negative charge down the profile (Table 9) from the adsorption of SO_4^{2-} and the increased ionic strength as would be expected from considerations of charge generation (Uehara and Gillman, 1981). In the upper

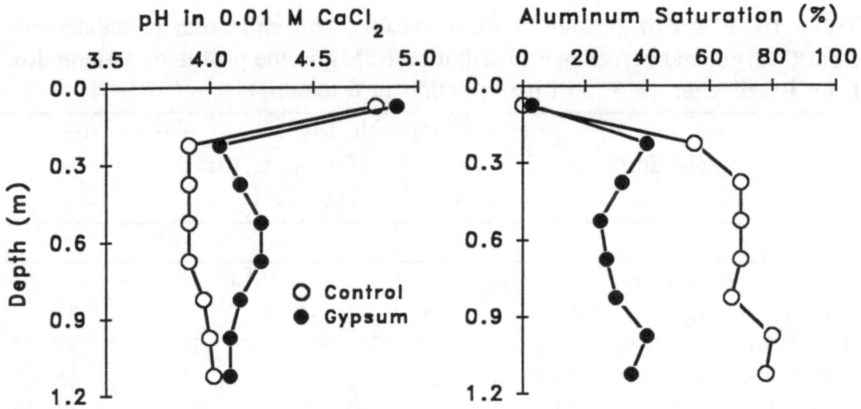

Figure 6. Effect of gypsum (6 Mg/ha) on soil pH and Al saturation in a Red Yellow Latosol (Xanthic Hapludox) profile sampled in 1983. (From Souza and Ritchey, 1986.)

part of the profile (0.15 m), the ionic strength due to the presence of large amounts of residual gypsum (van Raij and Peech, 1972; Gillman and Sumner, 1987) would be largely responsible for the increased charge whereas lower down the charge generated will be mainly due to sulfate adsorption which generates alkalinity as illustrated in Reaction 9. As a result, the pH of the subsoil increases and the Al^{3+} saturation decreases in the gypsum treatment (Figure 6). This effect has also been reported by Farina (1993) and Sumner et al. (1986). In all reported cases, gypsum application leads to an increase in subsoil exchangeable Ca levels (Shainberg et al., 1989; Sumner, 1990; Sumner, 1993). Both the reduction in Al and increase in Ca saturations contribute to improved root elongation and hence root proliferation in the subsoil (Noble and Sumner, 1988; Noble et al., 1988). In some soils, a release of Si into the solution has been observed after gypsum treatment (Lemus-Grob, 1985) which may reflect the lower stability of kaolinite relative to the aluminum hydroxy sulfate minerals at low pH as illustrated in Reaction 15. A frequent response to gypsum applications is the preferential movement and removal of Mg from the profile illustrated in Table 10 for a Typic Hapludox from Brazil and an Aquic Arenic Paleudult from Georgia. In the Paleudult, 45% of the Mg has been removed over a period of 4 years from the profile by 10 Mg gypsum/ha. In the Hapludox, the movement appears to be rapid with considerable redistribution taking place within 6 months of treatment. This is a function of the gypsum rate probably because this soil has a very low negative charge (Table 9) for its high clay content. On a Plinthic Paleudult, Farina (1993) found that Mg moved out of the topsoil as a result of gypsum treatment but tended to accumulate in the subsoil. Thus when gypsum is used on such soils, strategies must be developed to ensure the retention of adequate Mg levels in the profile for crop nutritional requirements. This can readily be achieved by applying Mg to the surface once most of the gypsum has

Table 10. Effect of gypsum on exchangeable Mg with depth in a Paleudult from Georgia and percentage distribution of Mg in the profile of a Hapludox from Brazil after 1.15 m of precipitation in 6 months

	Exchangeable Mg					
	Paleudult, Georgia		Hapludox, Brazil			
Depth	Gypsum rate (Mg ha^{-1})					
m	0	10	0	2	4	6
	$cmol_c\ kg^{-1}$		% of total			
0-0.15	0.25	0.15	75	54	39	33
0.15-0.30	0.23	0.10	4	16	14	10
0.30-0.45	0.10	0.04	2	9	16	15
0.45-0.60	0.06	0.03	3	3	9	13
0.60-0.75	0.08	0.04	4	4	6	7
0.75-0.90	0.09	0.05	4	5	5	8
0.90-1.05	0.10	0.09	4	4	6	6
1.05-1.20			4	4	5	6
Total	0.91	0.50				

(From Carvalho et al., 1986; Syed-Omar and Sumner, 1991.)

moved out of the topsoil (Syed-Omar and Sumner, 1991). Farina (1993) has found no evidence for the removal of Mg from soil profiles treated with gypsum and suggested that SO_4 generated CEC as illustrated in Table 9 may be responsible for arresting its movement. On the other hand, the movement of K is much more dependent on the type of soil. In the case of a number of Georgia Ultisols, Sumner (1990) and Syed-Omar and Sumner (1991) found virtually no movement except in a very sandy soil even 4 years after treatment with 10 Mg gypsum/ha. Similar results were obtained by Farina (1993). On a clay (Typic Hapludox) in a leaching column study, Souza and Ritchey (1986) found considerable movement after treatment with 6.9 Mg gypsum/ha (Figure 7). Liming the topsoil which would increase its ECEC reduced but did not eliminate the movement. Similar patterns of soil chemical behavior with respect to gypsum treatment have been reported by many workers but the increases in soil pH and net negative charge in the subsoil have not always been observed in the field (Gillman and Sumner, 1987; Shainberg et al., 1989; Sumner, 1990; Sumner, 1993). This may be due to soil heterogeneity masking the small differences between treatments.

When phosphogypsum is used, two additional benefits can accrue, namely, a yield response to the P content of the material (Shainberg et al., 1989; Meyer et al., 1991) and a further detoxification of subsoil Al^{3+} due to its F^- content by the formation of Al-F complexes which are less toxic (Alva et al., 1988; Alva and Sumner, 1988) and can be leached out of the profile (Oates and Caldwell, 1985; Keerthisinghe et al., 1991). In fact, the use of NaF proved to be highly

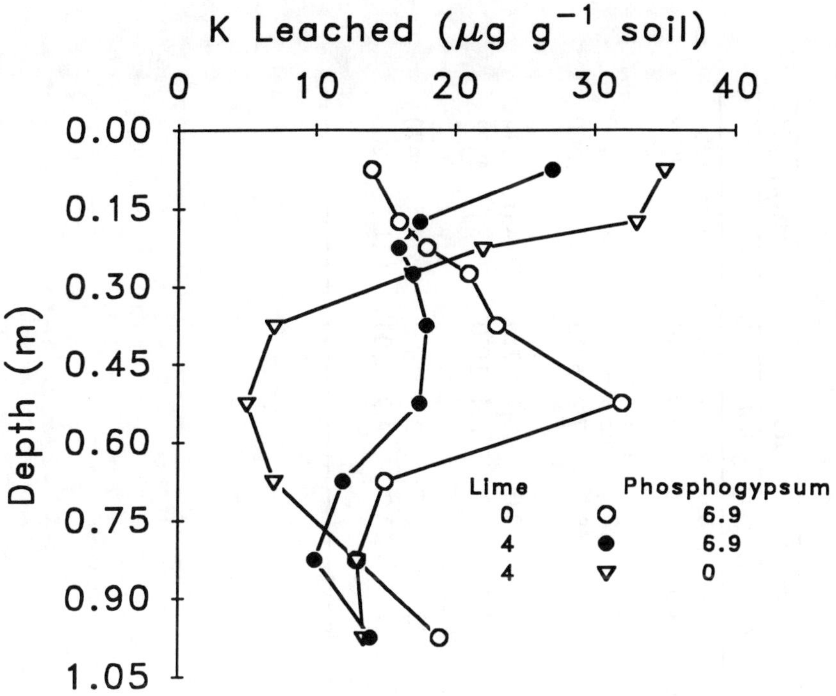

Figure 7. Influence of the application of gypsum, lime, and gypsum plus lime on K leached from a Xanthic Hapludox in a column leaching experiment. (From Ritchie et al., 1981.)

successful in reducing subsurface acidity but may cause health problems in livestock as the F content of the crop was increased (Keerthisinghe et al., 1991). The long term effect of the gypsum in single superphosphate in conjunction with lime has also been shown to have a beneficial effect on subsoil acidity (Ritchey et al., 1982; Ritchey, 1987).

The above changes in exchange chemistry which take place after gypsum application are also reflected in the chemistry of the soil solution (Table 11). Although application of gypsum usually increases the concentration of total Al in solution (Al_t), the activity of the toxic Al^{3+} ion is consistently reduced as a result of the formation of the ion pair, $AlSO_4^+$, which has been shown to be non-toxic or less toxic than Al^{3+} (Pavan et al., 1982; Alva et al., 1988; Noble et al., 1988). In addition, the activity of Ca^{2+} in solution is always increased.

The beneficial effects of gypsum on subsoil acidity are fairly longlasting. Sumner (1990) has demonstrated that the effect of a single application of gypsum (10 Mg/ha) was clearly measurable in terms of soil properties and yields of alfalfa after eight years. Farina (1993) showed that gypsum, at the same rate, was effective in increasing the yield of corn over nine years. When the cost of

Table 11. Effect of gypsum on the soil solution chemistry of various highly weathered soils speciated by the GEOCHEM program

		Concentration (μM)					Activity (μM)	
Soil	Treatment	Al_t	Al^{3+}	$AlSO_4^+$	AlOH[a]	Ca_t	Al^{3+}	Ca^{2+}
Oxisol subsoil, South Africa (Shainberg et al., 1989)	Control	8.9	3.4	0.0	2.7	140	2.6	107
	2 Mg ha^{-1} Gypsum	6.0	1.5	1.2	0.9	4610	0.4	2048
Inceptisol, Brazil (Pavan et al., 1987)	Control	96	36	10	45	100	21	190
	6.4 Mg ha^{-1} Gypsum	100	34	48	17	5210	9	2880
Ultisol subsoil, Georgia (Sumner et al., 1986)	Control	11	7.8	0.0	3.2	250	4.0	114
	10 Mg ha^{-1} Gypsum	14	7.3	3.3	3.4	1620	3.0	1123
Ultisol subsoil, South Africa (Buyeye and Fey, 1987)	Control	34	23	8.5	2.4	690	14	470
	3 Mg ha^{-1} Gypsum	38	21	12.2	4.9	2000	11	1200

[a]Sum of hydroxy species.
(From Sposito and Mattigod, 1979.)

Table 12. Effect of different calcium electrolytes at the same ionic strength on the pH of an Ultisol

Ionic strength	pH		
mol L^{-1}	$CaSO_4$	$CaCl_2$	ΔpH
0.0300	4.54	4.20	0.34
0.0140	4.72	4.44	0.28
0.0028	4.95	4.74	0.21
0.0014	5.08	4.91	0.17
0.0007	5.19	5.06	0.13

(From Sumner, 1990.)

gypsum is discounted over such a period of time, the economics of its use become very attractive. An economic analysis of the data from the experiment in Table 3 clearly demonstrated that the gypsum treatment was the most profitable (Farina, 1993). The persistence of these beneficial effects are likely to vary according to soil properties.

2. Possible Mechanisms for Amelioration

Before discussing the mechanisms involved, it is pertinent to review the general subsoil responses to gypsum application. Invariably, decreased exchangeable Al and increased Ca are found down the profile (Sumner, 1993). Frequently, especially when it is measured at equivalent ionic strength to eliminate the salt effect, pH increases after gypsum application (Table 12 and Figure 6) indicating a net release of alkalinity from the solid surfaces. In the soil solution, Ca^{2+} activity is increased and that of Al^{3+} reduced due to the formation of the non-toxic $AlSO_4^+$ ion pair and Al-F complexes (Table 11). The following mechanisms have been proposed to account for these observed effects:

a. Increased Subsoil Calcium

Root elongation will not take place in many crop plants in the presence of low levels of Ca in solution at the point of elongation (Hanson, 1984). Consequently, the improved Ca status brought about by gypsum entering the subsoil will overcome this limitation.

b. Ion Pairing

Due to the increased levels of SO_4^{2-} in subsoils, much of the soluble Al^{3+} becomes complexed as $AlSO_4^+$ which is non-toxic reducing the level of toxic

Al^{3+} in solution (Table 11). In systems in which F^- occurs in elevated concentrations (phosphogypsum and fluorogypsum), the formation of Al-F complexes will further detoxify the system by reducing the activity of Al^{3+}.

c. "Self-Liming" Effect

This was the first mechanism proposed for gypsum amelioration of acid subsoils (Reeve and Sumner, 1972) which involves the ligand exchange of SO_4^{2-} for OH on sesquioxide surfaces producing alkalinity (Table 12) which would neutralize soluble Al^{3+} as follows:

$$2[\mathrm{Fe,Al}]\begin{matrix}\mathrm{OH}\\ / \quad \backslash \\ \\ \backslash \quad / \\ \mathrm{OH}\end{matrix}[\mathrm{Fe,Al}] + Ca^{2+} + SO_4^{2-} \Rightarrow 2[\mathrm{Fe,Al}]\begin{matrix}SO_4^{-}\\ / \quad \backslash \\ \\ \backslash \quad / \\ \mathrm{OH}\end{matrix}[\mathrm{Fe,Al}] + Ca(OH)_2 \quad (9)$$

$$2\ Al^{3+} + 3\ Ca(OH)_2 \Rightarrow 2\ Al(OH)_3 + 3\ Ca^{2+} \quad (10)$$

d. Precipitation of Basic Aluminum Sulfate Minerals

Under acid conditions at elevated sulfate levels, basic aluminum sulfate minerals are known to precipitate (Adams and Rawajfih, 1977; Nordstrom, 1982) as follows:

$$3Al^{3+} + K^+ + 2Ca^{2+} + 2SO_4^{2-} + 3H_2O \Rightarrow \underset{\text{alunite}}{KAl_3(OH)_6(SO_4)_2} + 3\ H^+ + Ca^{2+} \quad (11)$$

$$4\ Al^{3+} = 10\ OH^- + SO_4^{2-} \Rightarrow \underset{\text{basaluminite}}{Al_4(OH)_{10}SO_4} \quad (12)$$

$$Al^{3+} + OH^- + SO_4^{2-} \Rightarrow \underset{\text{jurbanite}}{AlOHSO_4} \quad (13)$$

It is possible to generate alkalinity by reaction of gypsum with solid surfaces to form basic aluminum sulfates as follows:

$$Al(OH)_3 + CaSO_4 \Rightarrow AlOHSO_4 + Ca(OH)_2 \quad (14)$$

$$Al_2Si_2O_5(OH)_4 + CaSO_4 + 5\ H_2O \Rightarrow 2\ AlOHSO_4 + 2\ H_4SiO_4 + 2\ Ca(OH)_2 \quad (15)$$

$$2\ Al^{3+} + 3\ Ca(OH)_2 \Rightarrow Al(OH)_3 \quad (16)$$

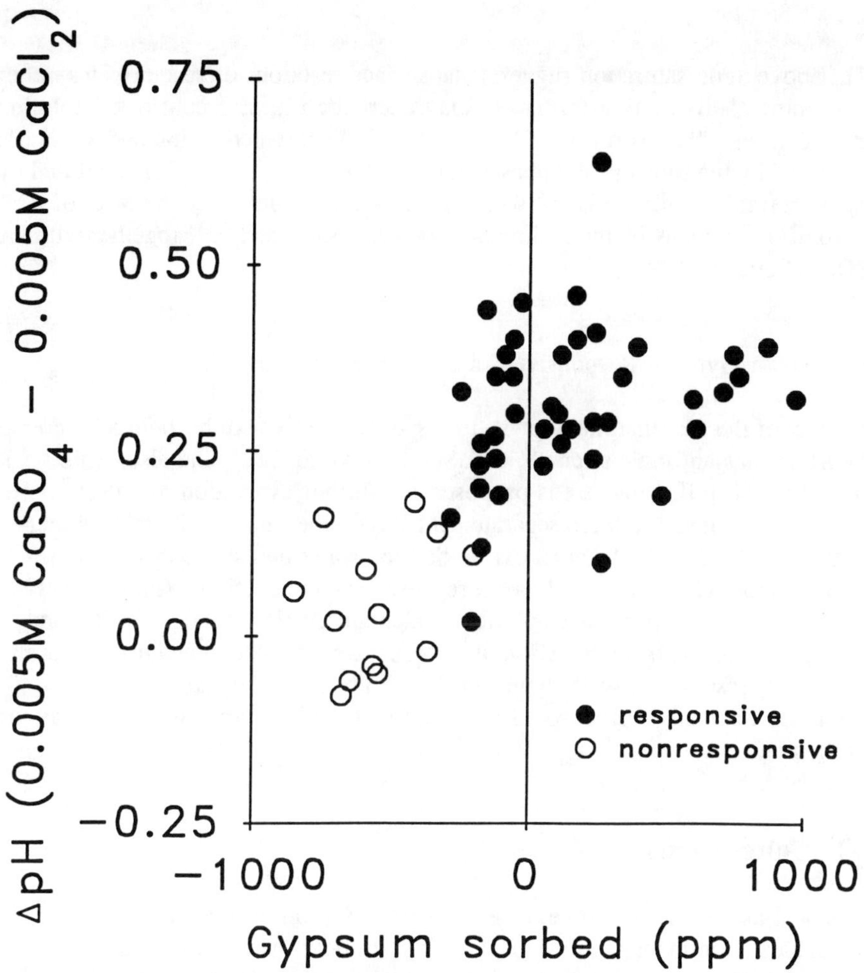

Figure 8. Relationship between ΔpH and gypsum sorbed for responsive and non-responsive subsoils. (From Sumner, 1990.)

Reaction (15) would explain the release of Si into solution observed in some soils. Although stability diagrams cannot be used to prove the existence of a solid phase, Sumner (1990) and Sumner et al. (1991) have shown, in some but not all soils, that conditions are appropriate for the formation of alunite and/or basaluminite. However, gypsum treated soils are often undersaturated with respect to all solid phases likely to be present.

e. Salt Sorption

The above undersaturation suggests that surface reactions on sesquioxides rather than mineral dissolution/formation may exert the ultimate control on solution composition. "Salt sorption" (Alva et al., 1991) in which cation and anion are adsorbed by the subsoil solid phases in equivalent amounts, is often exhibited by gypsum treated soils. This salt sorption can act as a control on the level of Al^{3+} in solution probably by the co-immobilization of Ca^{2+} and Al^{3+} together with the SO_4^{2-} (Sumner, 1990).

3. Test for Gypsum Responsive Soils

In view of the fact that many gypsum responsive soils, exhibit both an increase in pH at constant ionic strength in $CaSO_4$ relative to $CaCl_2$ and also exhibit salt sorption, the following test is proposed for further evaluation and use:

Take 3x10 g soil in three separate centrifuge tubes and add 25 ml of either 0.005M $CaSO_4$, 0.005M $CaCl_2$ or distilled water and shake intermittently overnight. Centrifuge and measure electrical conductivity (EC) and pH values in the supernatant solution. Calculate ΔpH ($CaSO_4$ - $CaCl_2$) and gypsum sorbed from the EC values. When ΔpH is plotted against gypsum sorbed, responsive soils should lie in the upper right quadrant.

Data for the soils tested so far in Figure 8 indicate that this test may be promising.

VI. Conclusions

Practical and economic strategies for the amelioration of subsoil acidity are now available. Continued liming of topsoils to near neutrality with inputs of acidifying fertilizers will, over a long period of time, provide sufficient excess of HCO_3^- capable of moving down the profile to neutralize acidity. However, if the topsoil pH is not maintained at a high enough level, this process will either not take place or will be very slow. This process can be accelerated by use of a voracious N feeding crop and adequate quantities of lime in conjunction with large inputs of acidifying nitrogenous fertilizers which together form $Ca(NO_3)_2$ which leaches into the subsoil where due to the differential uptake of NO_3^- and Ca^{2+} by roots, alkalinity is produced. In addition, various implements can be used to either incorporate lime economically in vertical bands or allow limed topsoil to fall down behind modified subsoilers to form vertical zones of ameliorated soil.

Gypsum, applied on the soil surface or incorporated into the topsoil, has proven to be an ameliorant which is highly efficient in reducing the deleterious effects of subsoil acidity on root growth and crop yields. When the gypsum front reaches the subsoil, amelioration is due to the increased supply of Ca^{2+} and the reduction in Al^{3+} by a number of mechanisms.

References

Abruna, F., J. Vicente-Chandler and R.W. Pearson. 1964. Effects of liming on yields and composition of heavily fertilized grasses and on soil properties under humid tropical conditions. *Soil Sci. Soc. Am. Proc.* 28:657-661.

Adams, F. 1981. Alleviating chemical toxicities: liming acid soils. p. 269-301. In: G.F. Arkin and H.M. Taylor (Eds.) *Modifying the root environment to reduce crop stress*. Am. Soc. Agr. Eng., St. Joseph. MI.

Adams, F. 1984. *Soil acidity and liming*, Ed. 2nd American Society of Agronomy, Madison, WI.

Adams, F. and P.J. Hathcock. 1984. Aluminum toxicity and calcium deficiency in acid subsoil horizons of two coastal plains soil series. *Soil Sci. Soc. Am. J.* 48:1305-1309.

Adams, F. and B.L. Moore. 1983. Chemical factors affecting root growth in subsoil horizons of Coastal Plain soils. *Soil Sci. Soc. Am. J.* 47:99-102.

Adams, F. and R.W. Pearson. 1969. Neutralizing soil acidity under bermudagrass sod. *Soil Sci. Soc. Am. Proc.* 33:737-742.

Adams, F. and Z. Rawajfih. 1977. Basaluminite and alunite: A possible cause of sulfate retention by acid soils. *Soil Sci. Soc. Am. J.* 41:686-692.

Adams, F., A.W. White and R.N. Dawson. 1967. Influence of lime sources and rates on Coastal bermudagrass production, soil profile reaction, exchangeable Ca and Mg. *Agron. J.* 59:147-149.

Alva, A.K. and M.E. Sumner. 1988. Effects of phosphogypsum or calcium sulfate on aluminon reactive aluminum in solutions at varying pH. *Comm. Soil Sci. Plant Anal.* 19:1715-1730.

Alva, A.K., M.E. Sumner and W.P. Miller. 1991. Salt absorption in gypsum amended acid soils. p. 93-97. In: R.J. Wright, V.C. Baligar and R.P. Murrmann (Eds.). *Plant-soil interactions at low pH.* Kluwer Academic Publishers, Netherlands.

Alva, A.K., M.E. Sumner and A.D. Noble. 1988. Alleviation of aluminum toxicity by phosphogypsum. *Comm. Soil Sci. Plant Anal.* 19:385-403.

Anderson, D.L. and J.G. Hendrick. 1983. Subsoil lime injector. *Soil Sci. Soc. Am. J.* 47:337-339.

Bataglia, O.C., C.E. Camargo, O.F. de Oliveira, V. Nagai and V.J. Ramos. 1985. Resposta à calagem de três cultivares de trigo com tolerância diferencial ao alumínio. *R. bras. Ci. Solo* 9:139-147.

Bolan, N.S., J.K. Syers and M.E. Sumner. 1991. Dissolution of various sources of gypsum in aqueous solutions and in soil. *J. Sci. Food Agric.* 57:527-541.

Bradford, J.M. and R.W. Blanchar. 1977. Profile modification of a Fagiudalf to increase crop production. *Soil Sci. Soc. Am. J.* 41:127-131.

Bromfield, S.M., R.W. Cumming, D.J. David and C.H. Williams. 1987. Long-term effects of incorporated lime and topdressed lime on the pH in the surface and subsurface of pasture soils. *Aust. J. Exp. Agric.* 27:533-538.

Brown, B.A. and R.I. Munsell. 1938. Soil acidity at various depths as influenced by the time since application, placement, and amount of limestone. *Soil Sci. Soc. Am. Proc.* 3:217-221.

Brown, B.A., R.I. Munsell, R.F. Holt and A.V. King. 1956. Soil reactions to various depths as influenced by time since application and amounts of limestone. *Soil Sci. Soc. Am. Proc.* 20:518-522.

Bruggenwert, M.G.M, T. Hiemstra and G.H. Bolt. 1991. Proton sinks in soil controlling soil acidification. p. 8-27. In: B. Ulrich and M.E. Sumner (Eds.). *Soil acidity*. Springer-Verlag, Berlin.

Buyeye, S.M. and M.V. Fey. 1987. Several explanations for the amelioration of acid, sesquioxidic soils by gypsum application. *Proc. Soil Sci. Soc. S. Afr. Conf.*

Caires, E.F. and C.A. Rosolem. 1991. Root growth of peanut cultivars and soil acidity. p. 239-243. In: R.J. Wright, V.C. Baligar and R.P. Murrmann (Eds.). *Plant-soil interactions at low pH*. Kluwer Academic Publishers, Dordrecht, The Netherlands.

Carvalho, L.J.C.B., R.L. Gomide, G.C. Rodriguez, D.M.G. Souza and E. de Freitas. 1986. Reposta do milho a aplicação de gesso e déficit hídrico em solo de Cerrados. *An. Sem. Uso. Fosfogesso Agricultura* 61-83.

Chaves, J.C.D., R. Figueiredo, F.B. Lima, C.E. Hirooka and M.A. Pavan. 1984. *Encontro sobre a cafeicultura Paranaense*. IAPAR, Londrina, Brazil.

Conyers, M.K. and B.J. Scott. 1989. The influence of surface incorporated lime on subsurface soil acidity. *Aust. J. Exp. Agric.* 29:201-207.

Coventry, D.R. 1991. The injection of slurries of lime, associated with deep tillage, to increase wheat production on soils with subsoil acidity. p. 437-445. In: R.J. Wright, V.C. Baligar and R.P. Murrmann (Eds.) *Plant-soil interactions at low pH*. Kluwer Academic Publishers, Dordrecht, The Netherlands.

Coventry, D.R., T.G. Reeves, H.D. Brooke, A. Ellington and W.J. Slattery. 1987. Increasing wheat yields in north-eastern Victoria by liming and deep ripping. *Aust. J. Exp. Agric.* 27:679-685.

Cumming, R.W. 1991. Long-term effects of lime in extensive pasture areas of Australia. p. 453-464. In: R.J. Wright, V.C. Baligar and R.P. Murrmann (Eds.). *Plant-soil interactions at low pH*. Kluwer Academic Publishers, Dordrecht, The Netherlands.

de Roo, H.C. 1956. Subsoiling, plowing, and deep placement of lime or fertilizer in one operation. *Agron. J.* 48:476-477.

Dematte, J.L.I. 1992. Aptidão agricola de solos e uso do gesso. *II Sem. Uso Gesso Agricultura* 307-324.

Doss, B.D., W.T. Dumas and Z.F. Lund. 1979. Depth of lime incorporation for correction of subsoil acidity. *Agron. J.* 41:541-544.

Ellington, A. 1986. Effects of deep ripping, direct drilling, gypsum and lime on soils, wheat growth and yield. *Soil Tillage Res.* 8:29-49.

Ellington, A., G.W. Ganning, and A.L. Avery. 1992. Lime slurry injection for acid subsoil. *Aust. Soil Acid. Res. Newslet.* 9:35.

Ernani, P.R., P.C. Cassol and G. Peruzo. 1992. Eficiência agronômica do gesso agrícola no Sul do Brasil. *II Sem. Uso Gesso Agricultura* 263-276.

Farina, M.P.W. 1993. Private communication.

Farina, M.P.W. and P. Channon. 1988a. Acid-subsoil amelioraton: I. A comparison of several mechanical procedures. *Soil Sci. Soc. Am. J.* 52:169-175.

Farina, M.P.W. and P. Channon. 1988b. Acid-subsoil amelioration: II. Gypsum effects on growth and subsoil chemical properties. *Soil Sci. Soc. Am. J.* 52:175-180.

Farina, M.P.W., M.E. Sumner, C.O. Plank and W.S. Letzsch. 1980. Aluminum toxicity in corn at near neutral soil pH levels. *J. Plant Nut.* 2:683-697.

Fox, R.L., N.V. Hue, R.C. Jones and R.S. Yost. 1991. Plant-soil interactions associated with acid, weathered soils. p. 197-204. In: R.J. Wright, V.C. Baligar and R.P. Murrmann (Eds.). *Plant-soil interactions at low pH.* Kluwer Academic Publishers, Dordrecht, The Netherlands.

Foy, C.D. 1992. Soil chemical factors limiting plant root growth. *Adv. Soil Sci.* 19:97-149.

Friesen, D.K., A.S.R. Juo and M.H. Miller. 1982. Residual value of lime and leaching of calcium in a kaolinitic ultisol in the high rainfall tropics. *Soil Sci. Soc. Am. J.* 46:1184-1189.

Gillman, G.P. 1991. The chemical properties of acid soils with emphasis on soils of the humid tropics. p. 3-14. In: R.J. Wright, V.C. Baligar, and R.P. Murrman (Eds.). *Plant-soil interactions at low pH.* Kluwer Academic Publishers, Dordrecht, The Netherlands.

Gillman, G.P. and M.E. Sumner. 1987. Surface charge characterization and soil solution composition of four soils from the Southern Piedmont in Georgia. *Soil Sci. Soc. Am. J.* 51:589-594.

Gillman, G.P. and E.A. Sumpter. 1986. Surface charge characteristics and lime requirements of soils derived from basaltic, granitic and metamorphic rocks in high- rainfall tropical Queensland. *Aust. J. Soil Res.* 24:173-192.

Goedert, W.J. 1983. Management of the Cerrado soils of Brazil: a review. *J.Soil Sci.* 34:405-428.

Gonzalez-Erico, E., E.J. Kamprath, G.C. Naderman and W.V. Soares. 1979. Effect of depth of lime incorporation on the growth of corn on an Oxisol of Central Brazil. *Soil Sci. Soc. Am. J.* 43:1155-1158.

Guimarães, P.T.G. 1992. O uso do gesso agrícola na cultura do cafeeiro. *II Sem. Uso Gesso Agricultura* 175-190.

Hammel, J.E., M.E. Sumner and H. Shahandeh. 1985. Effect of physical and chemical profile modification on soybean and corn production. *Soil Sci. Soc. Am. J.* 49:1508-1512.

Hanson, J.B. 1984. The functions of calcium in plant nutrition. p. 149-208. In: P.B. Tinker and A. Lanchli (Eds.). *Advances in Plant Nutrition.* Praeger, New York.

Helyar, K.R. 1991. The management of acid soils. p. 365-382. In: R.J. Wright, V.C. Baligar, and R.P. Murrman (Eds.) *Plant-soil interactions at low pH.* Kluwer Academic Publishers, Dortrecht, The Netherlands.

Helyar, K.R. and W.M. Porter. 1989. Soil acidification, its measurement and the processes involved. p. 61-101. In: A.D. Robson (Ed.). *Soil acidity and plant growth.* Academic Press, Australia.

Hern, J.L., A. Menser, R. Sidle, R.L. Wright and O.L. Bennett. 1982. The effects of organic complexing materials on ion movement and rooting environments in acid soils. *Agron. Abs.* 174.

Horsnell, L.J. 1985. The effect of superphosphate and lime on soil pH and on the establishment and growth of phalaris and lucerne. *Aust. J. Exp. Agric.* 25:149-156.

Howard, D.D. and F. Adams. 1965. Calcium requirement for penetration of subsoils by primary cotton roots. *Soil Sci. Soc. Am. Proc.* 29:558-562.

Hue, N.V., G.R. Craddock and F. Adams. 1986. Effects of organic acids on aluminum toxicity in subsoils. *Soil Sci. Soc. Am. J.* 50:28-34.

Jayawardane, N.S. and J. Blackwell. 1985. The effects of gypsum-enriched slots on moisture movement and aeration in an irrigated swelling clay. *Aust. J. Soil Res.* 23:481-492.

Joost, R.E. and C.S. Hoveland. 1986. Root development of Sericea lespedeza and alfalfa in acid soils. *Agron. J.* 78:711-714.

Juo, A.S.R. and J.C. Ballaux. 1977. Retention and leaching of nutrients in a limed Ultisol under cropping. *Soil Sci. Soc. Am. J.* 41:757-761.

Kamprath, E.J. and C.D. Foy. 1985. Lime-fertilizer-plant interactions in acid soils. p. 91-151. In: O.P. Englestad (Ed.) *Fertilizer technology and use.* Soil Science Society of America, Madison, WI.

Keerthisinghe, G., M.J. McLaughlin and J.R. Freney. 1991. Use of gypsum, phosphogypsum and fluoride to ameliorate subsurface acidity in a pasture soil. p. 509-517. In: R.J. Wright, V.C. Baligar, and R.P. Murrman (Eds.) *Plant-soil interactions at low pH.* Kluwer Academic Publishers, Dordrecht, The Netherlands.

Kirchhof, G., J. Blackwell and R.E. Smart. 1991. Growth of vineyard roots into segmentally ameliorated acidic subsoils. p. 447-452. In: R.J. Wright, V.C. Baligar, and R.P. Murrman (Eds.) *Plant-soil interactions at low pH.* Kluwer Academic Publishers, Dordrecht, The Netherlands.

Koch, D.W. and G.O. Estes. 1986. Liming rate and method in relation to forage establishment--crop and soil chemical responses. *Agron. J.* 78:567-571.

Kotze, W.A.G. and J. Deist. 1975. Amelioration of subsurface acidity by leaching of surface applied amendments. A laboratory study. *Agrochemophysica* 7:39-46.

Lathwell, D.J. and M. Peech. 1964. Interpretation of chemical soil tests. *Corn. Univ. Agric. Exper. Sta. Bul.* 995.

Lemus-Grob, F. 1985. *Effects and behavior of gypsum in an Ultisol*, M.S. Thesis, University of Georgia, Athens, GA.

Lobry de Bruyn, L.A. and A.J. Conacher. 1990. The role of termites and ants in soil modification: A review. *Aust. J. Soil Res.* 28:55-93.

Lund, Zane F. and B.D. Doss. 1980. Coastal bermudagrass yield and soil properties as affected by surface-applied dairy manure and its residue. *J. Environ. Qual.* 9:157-162.

Lutz, J.A. and G.D. Jones. 1971. Yield of wheat and soybeans and oil and protein content of soybeans as affected by fertility treatments and deep placement of limestone. *Agron. J.* 63:931-934.

Mahilum, B.C., R.L. Fox and J.A. Silva. 1970. Residual effects of liming volcanic ash soils in the humid tropics. *Soil Sci.* 109:102-109.

Malavolta, E. 1992. O gesso agrícola no ambiente e na nutricão da planta: Perguntas e respostas. *II Sem. Uso Gesso Agricultura* 41-66.

Manson, A.D. and M.V. Fey. 1989. Cation type and ionic strength effects on the solution composition of an acidic subsoil. *J. Soil Sci.* 40:577-583.

Marsh, G.H. and J.H. Grove. 1992. Surface and subsurface soil acidity: Soybean root response to sulfate-bearing spent lime. *Soil Sci. Soc. Am. J.* 56:1837-1842.

McKenzie, R.C. and M. Nyborg. 1984. Influence of subsoil acidity on root development and crop growth in soils of Alberta and northeastern British Columbia. *Can. J. Soil Sci.* 64:681-697.

McKenzie, R.C., D.C. Penney, L.W. Hodgins, B.S. Aulakh and H. Ukrainetz. 1988. The effects of liming on an ultisol in northern Zambia. *Comm. Soil Sci. Plant Anal.* 19:1355-1369.

McLay, C.D.A., G.S.P. Ritchie and W.M. Porter. 1992. Amelioration of subsoil acidity using surface-applied lime and gypsum in Western Australia. *Aust. Soil Acid. Res. Newslet.* 9:17.

Messick, D.L., M.M. Alley and L.W. Zelazny. 1984. Movement of calcium and magnesium in Ultisols from dolomitic limestone. *Soil Sci. Soc. Am. J.* 48:1096-1101.

Meyer, J., P.E.T. Turner and M.V. Fey. 1991. Interim evaluation of phospho gypsum as an ameliorant for soil acidity in sugarcane. *Proc. S. Afr. Sug. Tech. Assoc.* 65:41-46.

Neilsen, G.H., E. Hogue and B.G. Drought. 1981. The effects of surface-applied calcium on soil and mature Spartan apple trees. *Can. J. Soil Sci.* 61:295-302.

Noble, A.D. and M.E. Sumner. 1988. Calcium and Al interactions and soybean growth in nutrient solutions. *Comm. Soil Sci. Plant Anal.* 19:1119-1131.

Noble, A.D., M.E. Sumner and A.K. Alva. 1988. The pH dependency of aluminum phytotoxicity alleviation by calcium sulfate. *Soil Sci. Soc. Am. J.* 52:1398-1402.

Nordstrom, D.K. 1982. The effect of sulfate on aluminum concentrations in natural waters: some stability relations in the system Al_2O_3-SO_3-H_2O at 298 K. *Geochim. Cosmochim. Acta.* 46:681-692.

Oates, K.M. and A.G. Caldwell. 1985. Use of by-product gypsum to alleviate soil acidity. *Soil Sci. Soc. Am. J.* 49:915-918.

Pavan, M.A., F.T. Bingham and F.J. Peryea. 1987. Influence of calcium and magnesium salts on acid soil chemistry and calcium nutrition of apple. *Soil Sci. Soc. Am. J.* 51:1526-1530.

Pavan, M.A., F.T. Bingham and P.F. Pratt. 1982. Toxicity of aluminum to coffee in Ultisols and Oxisols amended with $CaCO_3$, $MgCO_3$, and $CaSO_4.2H_2O$. *Soil Sci. Soc. Am. J.* 46:1201-1207.

Pearson, R.W., F. Abruna and J. Vicente-Chandler. 1962. Effect of lime and nitrogen applications on downward movement of calcium and magnesium in two humid tropical soils of Puerto Rico. *Soil Sci.* 93:77-82.

Pearson, R.W., J. Childs and Z.F. Lund. 1973. Uniformity of limestone mixing in acid subsoil as a factor in cotton root penetration. *Soil Sci. Soc. Am. Proc.* 37:727-732.

Peoples, M.B., V.F. Burnett, A.M. Ridley, C.J. Smith, D.L. Garden, J.D. Williams, G. Keerthisinghe, T. James, and J.R. Freney. 1992. Evaluation of strategies to ameliorate subsurface acidity in pasture soils. *Aust. Soil Acid. Res. Newlet.* 9:36.

Percival, G.P., D. Josselyn and K.C. Beeson. 1955. Factors affecting the mirconutrient element content of some forages in New Hampshire. *New Hamp. Agric. Exp. Sta.* 93.

Perez-Escolar, R. and M.A. Lugo-Lopez. 1978. Effect of depth of lime application on yield of two corn hybrids grown on a typical ultisol of Puerto Rico. *J. Agric. Univ. Puerto Rico* 62:203-213.

Pleysier, J.L. and A.S.R. Juo. 1981. Leaching of fertilizer ions in a ultisol from the high rainfall tropics: Leaching through undisturbed soil columns. *Soil Sci. Soc. Am. J.* 45:754-760.

Pohlman, G.G. 1946. Effect of liming different soil layers on yield of alfalfa and on root development and nodulation. *Soil Sci.* 62:255-266.

Porter, W.M. and K.R. Helyar. 1992. Subsurface acidity constraints to agricultural production. *Natl. Work. Subsoil Constr. Root Grow., Australia.*

Quaggio, J.A. 1992. Reposta das culturas de milho e soja, a aplicação de calcário, gesso e movimento de ions em solos de Estado da São Paulo. *II Sem. Uso Gesso Agricultura* 241-262.

Rechcigl, J.E., R.B. Reneau and D.E. Starner. 1991. Alfalfa yields and quality as influenced by subsurface application of phosphorus, potassium and limestone. p. 485-490. In: R.J. Wright, V.C. Baligar and R.P. Murrmann (Eds.) *Plant-soil interactions at low pH*. Kluwer Academic Publishers, Dordrecht, The Netherlands.

Rechcigl, J.E., R.B. Reneau and D.E. Starner. 1985a. Effect of subsurface amendments and irrigation on alfalfa growth. *Agron. J.* 77:72-75.

Rechcigl, J.E., R.B. Reneau, Jr., and D.E. Starner. 1985b. Movement of subsurface fertilizer and limestone under irrigated and nonirrigated conditions. *Soil Sci.* 140:442-448.

Rechcigl, J.E., R.B. Reneau and D.E. Starner. 1985c. Movement of fertilizer and limestone in a Tatum clay loam. *Comm. Soil Sci. Plant Anal.* 16:427-438.

Rechcigl, J.E., D.D. Wolf, R.B. Reneau and W. Kroontje. 1985d. Influence of surface liming on yield and nutrient concentration of alfalfa established using no-tillage techniques. *Agron. J.* 77:956-959.

Reeve, N.G. and M.E. Sumner. 1972. Amelioration of subsoil acidity in Natal Oxisols by leaching of surface-applied amendments. *Agrochemophysica* 4:1-6.

Rena, A.B., E. Malavolta, M. Rocha and T. Yamada. 1986. *Cultura do cafeeiro: Fatores que afetam a productividade*. Associação Brasileira para Pesquisa da Potassa e do Fosfato, Piracicaba, SP, Brazil.

Reuss, J.O. 1991. The transfer of acidity from soils to surface waters. p. 203-217. In: B. Ulrich and M.E. Sumner (Eds.) *Soil acidity*. Springer--Verlag, Berlin.

Ridley, A.M., W.J. Slattery, K.R. Helyar and A. Cowling. 1990. The importance of the carbon cycle to acidification of a grazed annual pasture. *Aust. J. Exp. Agric.* 30:529-537.

Riley, W.R. 1991. *Yield, physiology and soil moisture extraction of alfalfa and Sericea lespedeza as affected by gypsum and irrigation*, Ph.D. Diss., University of Georgia, Athens, GA.

Ritchey, K.D. 1987. Efficient use of precipitation in humid regions. p. 327-336. In: L.L. Boersma, D.E. Elrick, R.B. Corey, H.H. Cheng, T.C. Tucker, and E.M. Rutteds (Eds.) *Future developments in soil science research*. Soil Sci. Soc. Amer., Madison, WI.

Ritchey, K.D. 1981. Calcium leaching to increase rooting depth in a Brazilian savannah Oxisol. *Agron. J.* 72:40-44.

Ritchey, K.D., J.E. Silva and U.F. Costa. 1982. Calcium deficiency in clayey B horizons of Savanna Oxisols. *Soil Sci.* 133:378-382.

Ritchie, G.S.P. 1989. The chemical behaviour of aluminium, hydrogen and manganese in acid soils. p. 1-60. In: A.D. Robson (Ed.) *Soil acidity and plant growth*. Academic Press, Sydney.

Ritter, W. and W.J. Sorrenson. 1985. *Cattle production on the highlands of Santa Catarina, Brazil*. TZ-Verlagsgesellschaft, Rossdorf, Germany.

Rowe, B.A. 1982. Effects of limestone on pasture yields and the pH of two krasnozems in north-western Tasmania. *Aust. J. Exp. Agric. Anim. Husb.* 22:100-105.

Saayman, D. and L. van Huyssteen. 1981. The efficiency of different methods of lime application during soil preparation. *S. Afr. J. Enol. Vitic.* 2:29-36.

Shainberg, I., M.E. Sumner, W.P. Miller, M.P.W. Farina, M.A. Pavan and M.V. Fey. 1989. Use of gypsum on soils: a review. *Adv.Soil Sci.* 9:1-111.

Shamshuddin, J., I. Che Fauziah and H.A.H. Sharifuddin. 1991. Effects of limestone and gypsum application to a Malaysian ultisol on soil solution composition and yields of maize and groundnut. p. 501-508. In: R.J. Wright, V.C. Baligar, and R.P. Murrman (Eds.). *Plant-soil interactions at low pH.* Kluwer Academic Publishers, Dordrecht, The Netherlands.

Souza, D.M.G., E. Lobato, K.D. Ritchey and T.A. Rein. 1992. Resposta de culturas anuais e leucena a gesso no Cerrado. *II Sem. Uso Gesso Agricultura* 277-306.

Souza, D.M.G. and K.D. Ritchey. 1986. Uso do gesso no solo de Cerrado. *An. Sem. Uso. Fosfogesso Agricultura* 119-144.

Sposito, G. 1989. *The chemistry of soils.* Oxford University Press, New York, NY.

Sposito, G. and Mattigod, S.V. 1979. *Geochem: a computer program for the calculation of chemical equilibria in soil solutions and other natural water systems*, Kearney Foundation of Soil Science, University of California, Riverside, CA.

Springett, J.A. 1983. Effect of five species of earthworm on some soil properties. *J. App. Ecol.* 20:865-872.

Stewart, R. and F.A. Wyatt. 1919. Limestone action on acid soils. *Ill. Agr. Exp. Sta. Bul.* 212:

Sumner, M.E. 1990. *Gypsum as an ameliorant for the subsoil acidity syndrome.* Florida Institute of Phosphate Research, Bartow, FL.

Sumner, M.E. 1991. Soil acidity control under the impact of industrial society. p. 517-541. In: G.H. Bolt (Ed.) *Interactions at the soil colloid - soil solution interface.* Kluwer Academic, Netherlands.

Sumner, M.E. 1993. Gypsum and acid soils: the world scene. *Adv.Agron.* (In Press).

Sumner, M.E., M.V. Fey and A.D. Noble. 1991. Nutrient status and toxicity problems in acid soils. p. 149-182. In: B. Ulrich and M.E. Sumner (Eds.) *Soil acidity.* Springer-Verlag, Berlin.

Sumner, M.E., H. Shahandeh, J. Bouton and J.E. Hammel. 1986. Amelioration of an acid soil profile though deep liming and surface application of gypsum. *Soil Sci. Soc. Am. J.* 50:1254-1258.

Sumner, M.E., L. West, and J. Leal. 1993. *Suelos de la agroindustria cafetalera del Guatemala: Region sur*, University of Georgia, Athens, GA.

Syed-Omar, S.R. and M.E. Sumner. 1991. Effect of gypsum on soil potassium and magnesium status and growth of alfalfa. *Comm. Soil Sci. Plant Anal.* 22:2017-2028.

Takow, J.A., M.D. Doumbia and L.R. Hossner. 1991. Acid soil profiles of the semiarid and subhumid tropics in Central and West Africa. p. 313-320. In: R.J. Wright, V.C. Baligar, and R.P. Murrman (Eds.) *Plant-soil interactions at low pH*. Kluwer Academic Publishers, Dordrecht, The Netherlands.

Thomas, G.W. and W.L. Hargrove. 1984. The chemistry of soil acidity. p. 3-56. In: F. Adams (Ed.) *Soil acidity and liming*. American Society of Agronomy, Madison.

U.S.Survey Staff, 1975. *Soil Taxonomy; A Basic System of Soil Classification for Making and Interpreting Soil Surveys*, United States Department of Agriculture SCS, Washington D.C..

Uehara, G. and Gillman, G. 1981. *The mineralogy, chemistry, and physics of tropical soils with variable charge clays*, Westview Press Inc., Boulder, CO.

Ulrich, B. and Sumner, M.E. 1991. *Soil acidity*, Springer-Verlag, Berlin.

van Breemen, N. 1991. Soil acidification and alkalinization. p. 1-7. In: B. Ulrich and M.E. Sumner (Eds.) *Soil acidity*. Springer-Verlag, Berlin.

van Raij, B. and M. Peech. 1972. Electrochemical properties of some Oxisols and Alfisols of the tropics. *Soil Sci. Soc. Am. Proc.* 36:587-593.

Vitti, G.C., J.A. Mazza, H.S. Pereira and J.L.I. Dematte. 1992. Resultados esperimentais do uso de gesso na agricultura. *II Sem. Uso Gesso Agricultura* 191-224.

Weir, C.C. 1975. Effect of lime and nitrogen application on citrus yields and on the downward movement of calcium and magnesium in a soil. *Trop. Agric. (Trinidad)* 51:230-234.

White, P.F. and A.D. Robson. 1989. Effect of soil pH and texture on the growth and nodulation of lupins. *Aust. J. Agric. Res.* 40:63-73.

Wright, R.J., J.L. Hern, V.C. Baligar and O.L. Bennett. 1985. The effect of surface applied soil amendments on barley root growth in an acid subsoil. *Comm. Soil Sci. Plant Anal.* 16:179-192.

Stress Transmission and Recompaction in Tilled and Segmently Disturbed Subsoils under Trafficking

R. Horn

I. Introduction

Soil compaction as a world wide problem is discussed very intensely in soil science literature (e.g., Scheffer and Schachtschabel, 1992; Hartge and Horn, 1991). The development and use of heavy agricultural machinery and its use through the years has changed the soil physical, chemical, and biological properties to conditions less favorable for plant growth. This is true, because the maximum mass of various agricultural machines used during agricultural operations has increased by a factor of 3-4 during the last 3 decades and the number of field operations can reach up to 14 treatments a year. Thus, the maximum acceptable loading and/or the degree of compaction of arable and forest soils have to be carefully considered in order to protect the soil as a sustainable system and to use it at its optimal potential.

In recent years, informations about the positive and negative effects of soil compaction on the environment have been reviewed in relation to its importance for several types of ecosystems in the various regions of the world and under various climatic and site conditions (ISTRO, 1991).

1-56670-020-5/95/$0.00+.50

From the agricultural viewpoint the reduction in crop yields due to soil compaction is the most important effect. This is often connected with reduced water and air flow into surface and subsurface layers and restricted root growth. Another important effect of soil compaction is increased runoff and soil erosion (Grath, 1993). Hakansson et al. (1988) described the long lasting effect of soil compaction on yield reduction, as well as on the increased sensitivity to drought or to soil wetness due to a single intensive wheeling event at the "wrong time". In the eastern states of Germany (former German Democratic Republic) and other east European countries, farmers have used very large and heavy agricultural machinery in order to finish the different tillage and agronomic operations in time. However, because of the use of these heavy machines under unfavorable soil conditions, intensive soil compaction has occurred in soil layers far below the plow pan. For instance in the state of Thuringia about 30% of the arable soils show significant soil structure damage which result in a yield decrease of up to 30% (Werner et al., 1992). Field studies show the high susceptability of the agriculturally very productive mollisols and luvisols derived from loess to soil compaction, even to great depths.

Various tillage systems such as conventional or conservational systems which use biological processes (e.g., earthworms or loosening by root growth) can be applied to loosen compacted soil horizons. However, the initial improvement of the soil is more effective if done by mechanical methods followed thereafter by biological processes like stabilization by organic-mineralic bondings.

Eck and Unger (1985) and Jayawardane et al. (1994) pointed out that the type of soil loosening practices required depends on the specific agronomic problems at the site. For instance, improving sodic subsoils require soil loosening to depth, combined with addition of gypsum to increase the macropore stability of the loosened layer. Improving soils with subsoil acidity requires uniform mixing of lime to depth (FBW 1989; Horn 1994). The crop responses to such soil loosening practices, however, varies with the variations in climate during the cropping seasons. Once the need for soil loosening is determined and the most effective and long lasting technique of soil loosening is chosen, subsequent traffic management required to minimize soil recompaction needs to be selected. Achieving long term soil improvements through soil loosening is dependent on the physical and physico-chemical properties of the loosened soil and control of subsequent trafficking. Trafficking, especially under wet conditions where the soil strength of the loosened soil is low, can lead to marked soil compaction.

II. Soil Loosening Techniques and Machinery

A detailed review of the problems of reduced water movement, root growth and crop yields on compacted soils with fragipans, duripans and claypans is provided by Eck and Unger (1985). They examined the effectiveness of soil loosening practices such as plowing, ripping, chiseling and trenching in increasing water movement and soil aeration. Crop responses to soil loosening, however, varies widely according to seasonal climatic conditions. (Jayawardane et al., 1994). Studies on conservation tillage practices show, that excessive surface soil disruption by soil loosening can cause reduced water and air flow into the subsoil, creating adverse soil conditions (Horn et al., 1993: Horn and Lebert, 1992).

A wide range of soil loosening techniques have been used on agricultural lands in European countries, over long periods of time. In 1837, an implement refered to as the "Miner" was used (cited in Schulte Karring, 1988). The loosening depths of plowing was about 30 cm and thereby far below the compacted plowpan layer. In the following decades several attempts had been undertaken to improve soil properties by loosening, including those of Read (Reads Subsoil Pulverizer, 1841), and Tweedalle (Tweedalle's Subsoil Trench Plough, 1850).

Between 1860 and 1925, deep loosening techniques in general became more powerful due to the application of steam engines. From 1935 on, rigid deep loosening tines were added to the normal plows. In the early 1960s, the rigid tool was replaced by dynamic devices which were called "Hubschwenk (1968), Wippschar (1970) or Stechhublockerer (1975)". In order to increase the traction forces and to further reduce the contact area pressure of the engines very heavy crawlers with very wide and long chains were used for soil reloosening operations. However, at a given contact area pressure, stresses are transmitted to deeper depths with increase in contact area. Consequently the bigger mass of the crawler itself, even with a greater contact area, may induce a deeper and more pronounced stress distribution and hence soil compaction (Horn et al., 1987). Therefore, farmers had to apply deeper working soil reloosening technique, which again require the use of more powerful and heavier machines. Nowadays the crawlers required for such soil reloosening have a mass of up to 30 Mg and a contact area of up to 5 m^2 (Lebert, 1991).

Other loosening techniques like exploding dynamite or applying air pressure to loosen the soil by lifting, and to create further cracks were also used. These anthropogenic cracks were filled with perlite or other kinds of incomposable material in order to preserve the reloosened soil conditions (Schulte Karring, 1991). In general both techniques are only economically acceptable where only small soil volumes have to be improved (as in vineyards or in horticulture). It also requires soils dry enough to create brittle failure during amelioration.

Recently, a new reloosening technique called the "Abbruchlockerung" (Schulte Karring, 1990; 1991) has been developed in order to improve the effectiveness of soil loosening, and to minimize the possibility of amelioration failure, especially under wetter and less stable site conditions. This technique reduces the intensive shearing and kneading of soil during the amelioration process. Comparable to the mixing technique of the spading machine, the soil is turned around and becomes more or less homogenized. The loosening device itself pushes the tractor forward during the soil amelioration.

Thus 3 different types of soil loosening techniques can be distinguished:
- deep furrow plowing and subsoiling,
- partial (point) reloosening, and
- complete mixing or loosening soil layers.

All these techniques destroy the "natural" soil structure completely during the reloosening event. Use of these techniques result in completely loosened soil with low mechanical strength, which is susceptible to intensive soil compaction if wheeled as before.

Instead of a complete soil reloosening of the entire soil volume, Jayawardane and Blackwell (1985) described a new slotting technique which results only in a segmental soil amelioration. This results in loosening of only narrow sections of the soil profile. It is comparable with the ditch plowing technique of Reich et al. (1985). Both techniques combine the advantages of having an ameliorated soil volume for better plant growth and improved water and air infiltration, with the presence of a nonloosened strong soil volume to carry the load of tractors.

In the following sections the important soil properties and stress distribution patterns under wheeling which influence the extent of soil recompaction are discussed.

III. Mechanical Strength and Stress Distribution during Wheeling

A. " Natural Soils"

1. Mechanical Strength

Drescher et al. (1988), Larson et al. (1989), and Soane and van Ouwerkerk (1994) showed that depending on the internal soil parameters as well as on external factors, each soil horizon has a well defined mechanical strength value which can be quantified by its precompression stress (Horn 1981). In general, soils are the stronger, the more intensively aggregated they are, the smaller the pore water pressure; the higher the organic carbon content, kind of organic acids (Hempfling et al., 1990) and the higher the base saturation of the exchange complex at a given clay mineralogy. In soils with markedly differentiated soil horizons, the differences in the precompression stress values of soil layers at a given pore water pressure are greater (e.g. -6 kPa) (Figure 1). In the luvisol derived from loess, the clay depleted Al-horizon is weaker compared to the clay enriched Bt horizon, where the highest precompression stress values occur. The crumbly structured A-horizons of the mollisol derived from loess are fairly strong in spite of their low bulk density values. The high strength of the well aggregated clayey horizon of the vertisol derived from mesozoic clay could be related to the very high water content at pore water pressure of -6kPa because the strength increasing menisci forces can react at a greater pore surface area (X

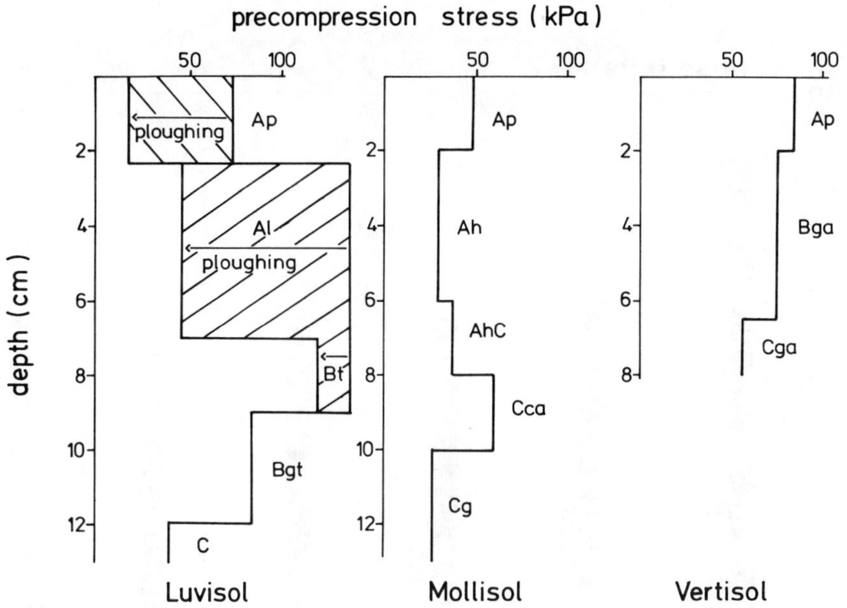

Figure 1. Precompression stress values of various soil types at a pore water pressure of -6kPa. The marked areas (Ap and Al horizons) of the Luvisol derived from loess show alterations of soil strength due to plowing. The Luvisol as well as the Mollisol are derived from loess, while the Vertisol is derived from mesozoic material.

factor of the effective stress equation is ~ 1 at this pore water pressure (Horn, 1981; Horn, 1993; Burger et al., 1988). No-tilled soils are always much stronger compared to the conventionally annually plowed soils, due to a better rearrangement of soil particles and a more stress equilibrated pore system. An ancient plow pan in such originally plowed luvisol still exists after more than 2 decades in spite of an intensive increase in earthworm activity and number (Horn, 1986). Hence, such soils need to be mechanically loosened before adoption of conservation practices.

2. Stress Distribution

Each stress, applied at the soil surface or on the plow pan during wheeling or trampling, is transmitted three-dimensionally in the soil volume. The higher the water saturation and total amount of water per soil volume, the more will the stresses be transmitted hydrostatically. In the drier and/or stronger soils the stress distribution is distributed more horizontally (Horn, 1981).

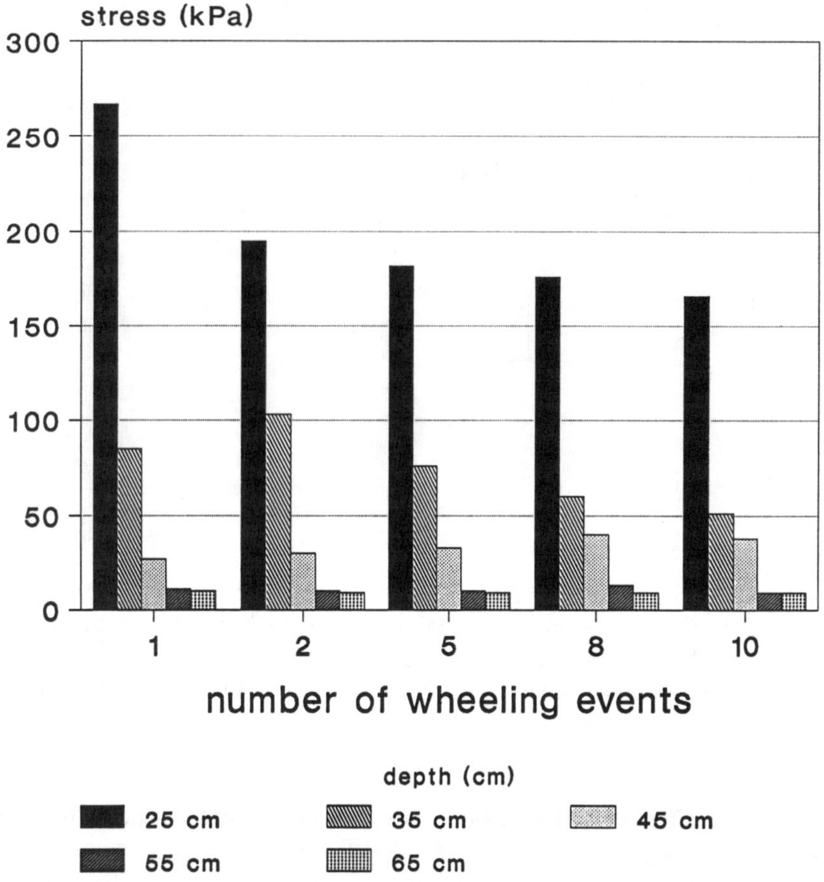

Figure 2. The effect of number of wheeling events on stress distribution and kind of stresses in a conventionally plowed gleyic Luvisol derived from glacial till.

As the shear strength of a soil layer is exceeded by the applied load the soil volume will be further compressed until a new stress dependent equilibrium between the external stresses and the internal soil strength is reached. In general, the registration of stresses in soils requires the application of Stress-State Transducers (SST) in order to predict the stress distribution and kind of stresses (Nichols et al., 1987; Horn et al., 1992). Figure 2 informs about the vertical stress in an undisturbed gleyic luvisol from glacial till as a function of the number of wheeling events. During the first wheeling at constant speed most of the stress is attenuated in the topsoil. The vertical stress dominates while the shear stresses are comparably small. During repeated wheeling, however, the vertical stresses in the top soil horizon seem to decrease due to the formation of a stronger platy structure during wheeling. The plate itself becomes strong enough to carry the load and to behave more or less elastically during the subsequent loading events. However, the deeper soil layers with low natural soil strength will now become more stressed, sheared, and consequently compacted during the following wheeling events. This can be expected from the higher normal and shear stress values during the consecutive wheeling events. Such additional soil compaction by shearing can also be observed as the result of slip intensity affected shear forces. If soils are wheeled with increasing slip percentage, the octohedral shear stresses increase and the mean normal stress and the horizontal principal stress decline.The vertical principal stress will also increase (Figure 3). Consequently, the soil becomes strongly overconsolidated to depths until a dynamic new internal soil strength equilibrium is obtained (Horn and Dexter, 1994).

B. Completely Mixed Soils

1. Mechanical Strength

The original soil structure dependent soil strength of undisturbed soil is reduced to the texture dependent values, when soils are either mixed completely by loosening or during seedbed preparation. Similar effects occur when soil material is refilled during landscape reconstruction in the brown coal mining excavation areas in Germany (Dumbeck, 1991) and in landforming of vineyards. Depending on the soil transportation system used, the refilled soil material may regain some strength, compared to the texture dependent strength values due to the formation of anthropogenic and very dense spherical aggregates (Lebert, 1991).

The decline in soil strength due to intensive mixing can also be shown (Figure 4) from the changes in the precompression stress data of forest soils in Germany. The original precompression stress distribution in a podsolic gleyic luvisol show a smaller strength value in the top Ah/Aeh horizon due to a low bulk density followed by a high value in the iron /aluminum and humus enriched Bsh horizon. The deeper soil layers are affected by gleyic processes. When these soils are completely mixed up to a depth of 80 cm by deep plowing or by chiseling in order to "ameliorate " the site just before reforestation, the original soil strength declines intensely. R defines the Rigosol properties, whereby RAp2 (org) refers to the soil layers with an enormous amount of organic material plowed in from the top, while RAp2 (min) refers to the strength of the mineral soil (Figure 4).

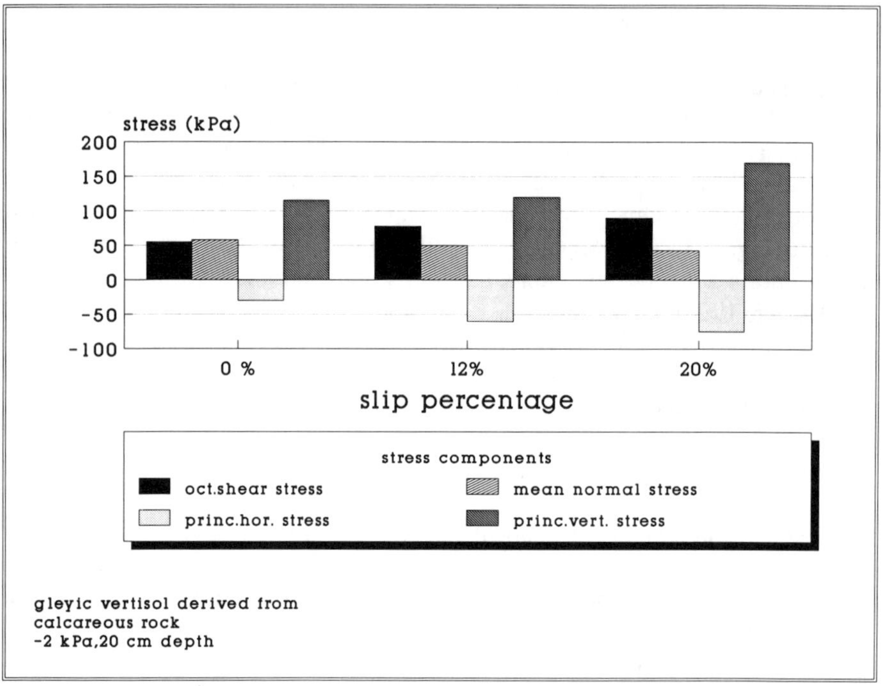

Figure 3. Effect of slip percentage on octahedral shear stress, mean normal stress, and the vertical as well as horizontal principal stresses.

2. Stress Distribution

In homogenized soils too, stresses are transmitted to depth three dimensionally. However, the perpendicular stress propagation dominates during wheeling. The stress equipotential lines are therefore more vertically oriented as compared to those for the well structured soils. The pattern of the stress equipotential lines can be described by the values of the concentration factor which range between 2 and 9 for very strong and soft soil conditions, respectively (according to Newmark, 1948, cited in Horn, 1993). At a given pore water pressure value the concentration factor values are smaller for well aggregated soils and higher for

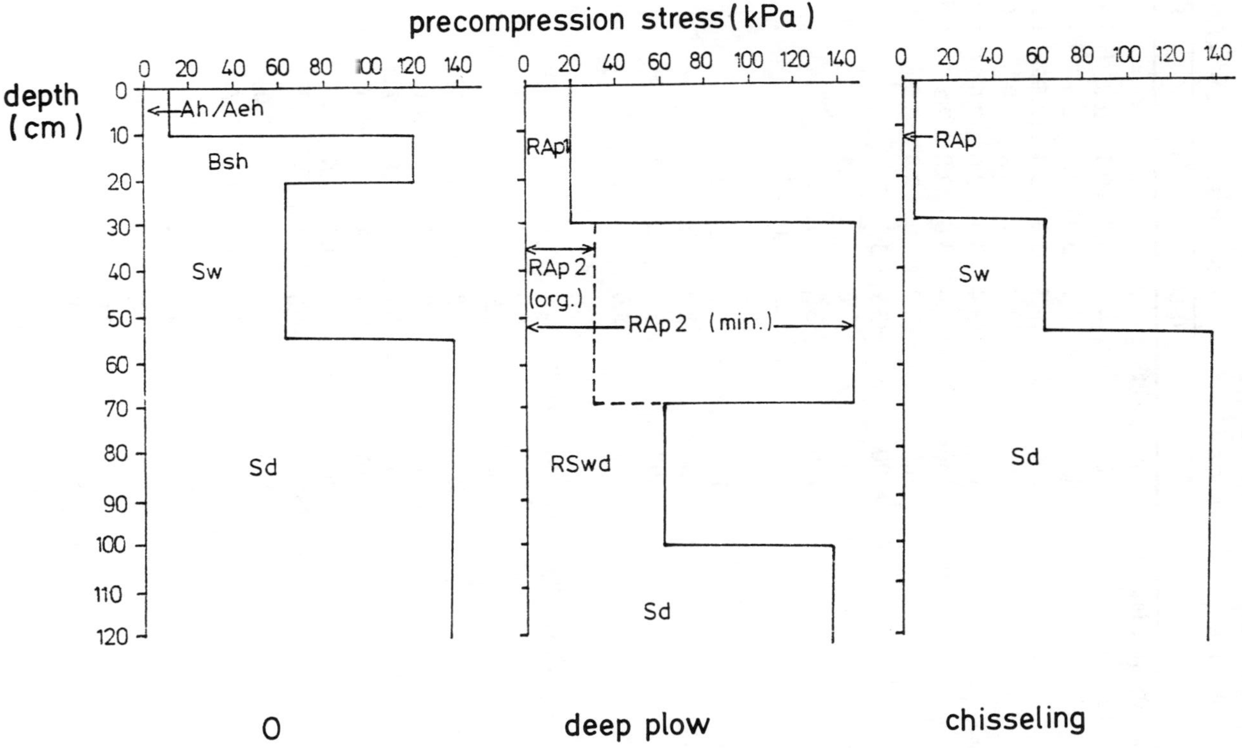

Figure 4. Changes in the precompression stress values in a podzolic gleyic luvisol under a forest due to amelioration by of chiseling, deep plowing and mixing with a pulling crawler.

Table 1. Strength dependent concentration factor values in silty and clayey soils at an approximate pore water pressure of -30 k Pa

Radius (cm) of the tire contact area	σ_0 (kPa) at the upper surface of each soil horizon[a]	Precompression stress (kPa)				
		0-60	-100	-140	-180	>180
Silt, sandy silt, loamy silt, clayey silt						
-10	100	4.1	3.7	2.1	2.0	1.1
	150	4.3	3.8	3.1	2.3	1.7
	200	4.5	3.9	3.4	2.5	1.8
	>200	4.7	4.2	3.8	2.9	2.0
10.1 - 15	100	3.4	3.3	2.6	2.6	1.9
	150	3.6	3.0	2.8	2.5	1.9
	200	3.7	3.0	2.9	2.8	2.0
	>200	3.9	3.7	-	-	-
15.1 - 20	100	3.0		2.4		
	150	3.3	2.7	2.6		
	200	3.5	3.0	2.8		
	>200	3.7	3.1			
20.1 - 25	100	2.5	2.3	2.1		
	150	2.9	2.7	2.5		
	200	3.4	3.0	2.7		
	>200	3.6		2.9		
Clay, sandy clay, silty clay, loamy clay						
-10	100	4.0	3.7	2.9	2.8	
	150	4.7	4.0	3.3	2.8	
	200	4.8	4.3	3.6	3.0	
	>200	4.9	4.5	3.9	3.1	
10.1 - 15	100	4.1	3.6	2.8	2.7	
	150	4.3	3.8	3.3	3.0	
	200	4.5	3.9	3.6	-	
	>200	4.8	4.4	3.8	-	
15.1 - 20	100	4.0	3.6	3.5	3.0	
	150	4.2	3.7	3.3	2.7	
	200	4.3	3.9	-	-	
	>200	4.5	4.0	-	-	
20.1 - 25	100	4.0	3.5	3.0	2.8	
	150	4.0	3.5	-	-	
	200	3.9	3.5	-	-	
	>200	3.7	3.6	-	-	

[a]Where σ_0 (kPa) is the applied stress.

loosened horizons with very small precompression stress values. Table 1 provides the precompression stress dependent concentration factor values for silty and clayey soils of various internal soil strength, at given applied stress, and for various tire contact areas (defined as their circle radius). The smaller precompression stress (i.e. the internal soil strength) of loosened soil leads to higher concentration factor values at a given applied stress. The greater the contact area of the tire, the deeper will the stresses be transmitted at a given contact area pressure.

Furthermore, wheeling of weak soil results in the creation of higher octahedral shear stresses at given normal stresses. Thus, a deeper and more pronounced soil recompaction occurs by shear induced particle rearrangement (Horn et al., 1994a).

The critical stress state can be also derived from the comparison of measured octahedral shear stress values and mean normal stress data during wheeling with the corresponding data defined by the Mohr Coulomb failure line for comparable situations. If the measured stresses exceed the defined failure criterion, soil structure must collapse until a new stress equilibrium is reached. The rearrangement of soil particles under the new stress system can be either induced by a vertical displacement and increase in bulk density or by a volumetric particle displacement which can even result in smaller bulk density values but an increased tortuosity of the pore system (Horn et al., 1994b).

IV. Stress Distribution in Soils during Total Soil Reloosening

As long as the applied external stress during wheeling is smaller than the soil strength at any depth the structure of the soil will not change. Thus, during tractor wheeling at the soil surface of an overconsolidated luvisol derived from loess the vertical stresses recorded in the Ap/Al/Bt/C horizons were smaller than the precompression stress values of 140/200/80/60 kPa (Figure 5) for the same horizons (Semmel, 1993). However during deep plowing, the deeper soil layers can be stressed by wheeling at a depth of 30 - 35 cm occurring on top of the plow pan. Thus, the high stress values of 140 and 80 kPa (Figure 6) in the deeper Bt and C horizons respectively exceed by far the internal soil strength at the given pore water pressure and induce a further soil compaction.

If arable soils are ameliorated by very heavy and powerful machines such as heavy crawlers further soil compaction processes could occur at depth. Thus, the principal vertical stresses, measured up to 80 cm depth during the reloosening of a luvisol (derived from loess), exceeded the internal soil strength (Figure 7). Consequently, this soil will be further compacted instead of becoming reloosened.

Such unexpected additional soil compaction during soil amelioration by deep loosening can be also very frequently seen in reconstructed refilled soils in the brown coal mining excavation area in Germany. After three years of lucerne growth for reaggregation purposes the maximum stresses during wheeling and

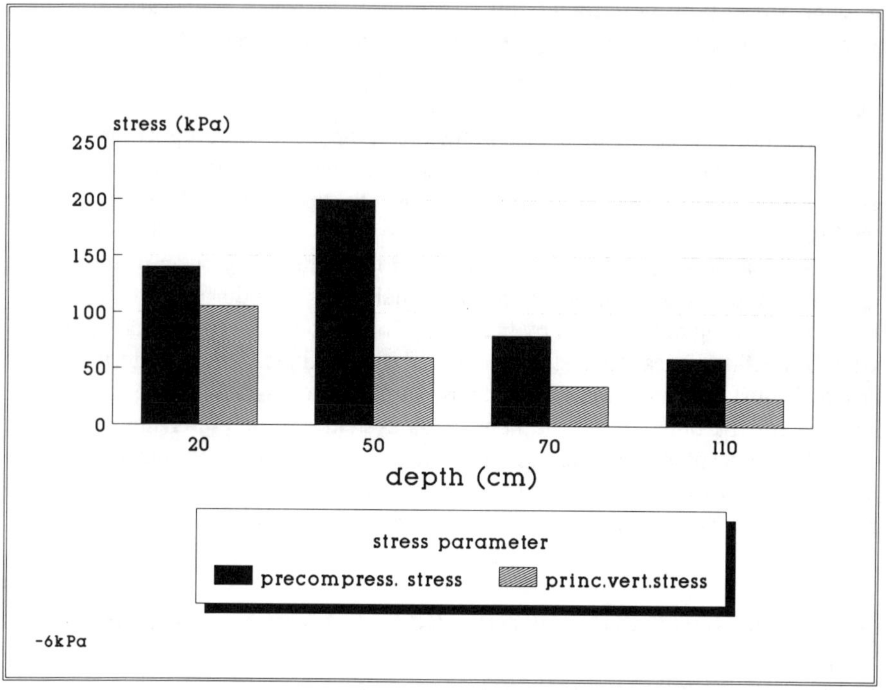

Figure 5. Vertical stresses in comparison with the soil strength in an undisturbed and overconsolidated Luvisol derived from loess during wheeling.

Table 2. Precompression stress data of the refilled soil (pararendzina derived from loess) at different depths at a pore water pressure of -6 kPa

Depth (cm)	Precompression stress (kPa)
0 - 15	20
15 - 40	70
40 - 80	40
80 - 120	35

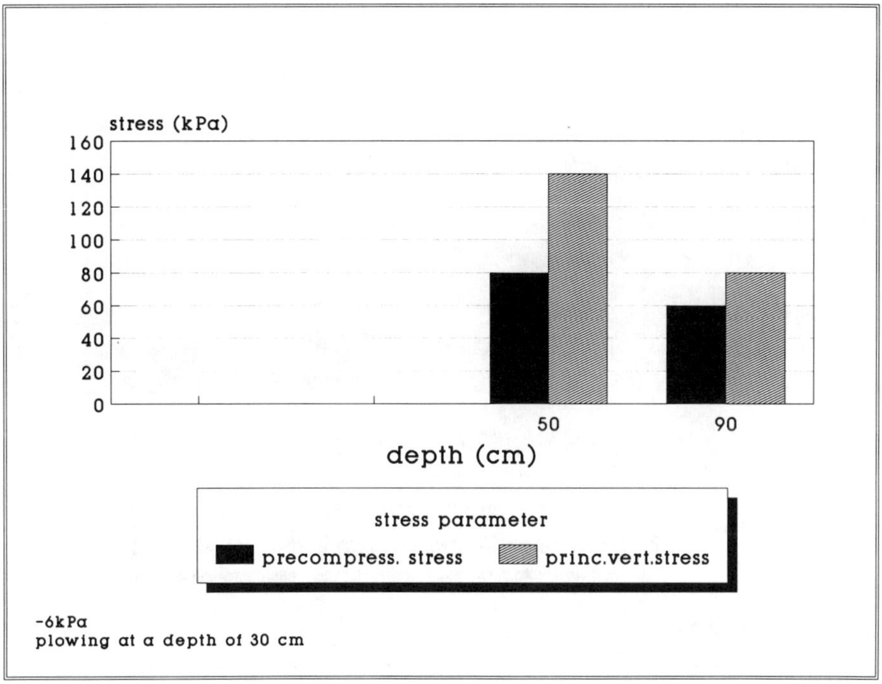

Figure 6. Vertical stresses in comparison with the soil strength in an undisturbed and overconsolidated Luvisol derived from loess during plowing at a depth of 30 cm, i.e., in the plow pan.

deep loosening with the Stechhublockerer were determined using the Stress State Transducer sensors at different depths (Baum, 1993). Compared to the corresponding precompression stress data for the various depths (Table 2), the stresses below the crawler chains were always greater when compared with the original soil strength. Even at a depth of 105 cm very high octahedral shear stresses (51kPa) and very high principal vertical stresses were recorded (Figure 8).

Furthermore, the stresses determined both between the reloosening tools and directly below them also revealed a significant increase, where the octahedral shear stresses as well as the principal vertical stresses dominated up to deeper

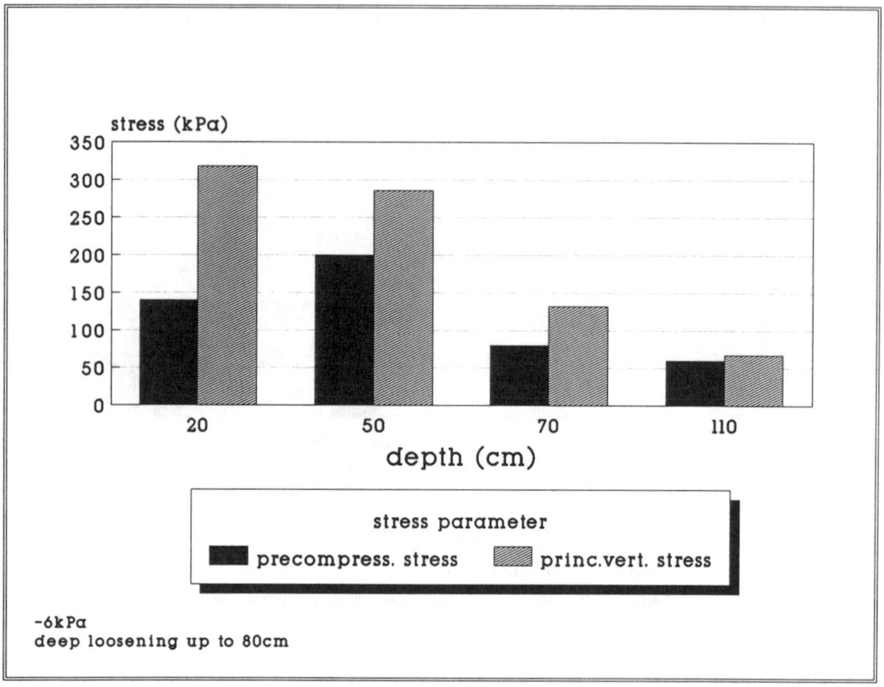

Figure 7. Vertical stresses in comparison with the soil strength in an undisturbed and overconsolidated Luvisol derived from loess during reloosening up to a depth of 80 cm.

depths (Figure 9). These high shear stresses may be responsible for intensive soil structure deterioration and in combination with a higher water content will lead to soil kneading and soil homogenization. This would result in a complete loss of soil strength.

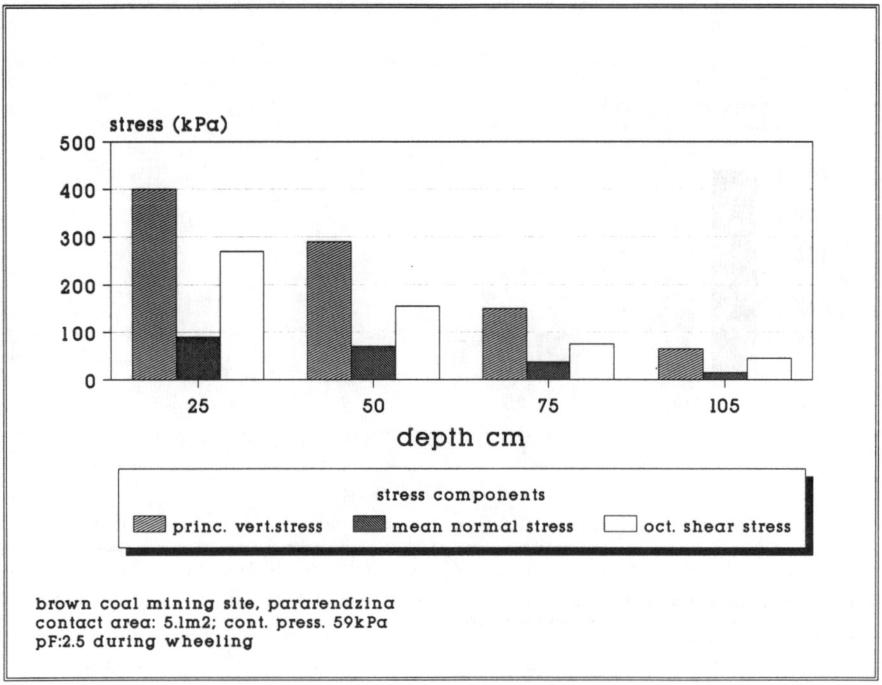

Figure 8. Stress distribution in a refilled pararendzina derived from loess during surface leveling with a crawler (brown coal mining area, Rheinbraun).

V. Stress Distribution in Segmentally Reloosened Soils

As shown in the previous sections, the complete reloosening of the soil resulted in a marked decline in soil strength which could lead to further soil compaction with wheeling. Because the wheeling intensity and the mass of the agricultural machinery has been increased in the last 3 decades, it is necessary to develop methods of soil loosening which can withstand such trafficking requirements.

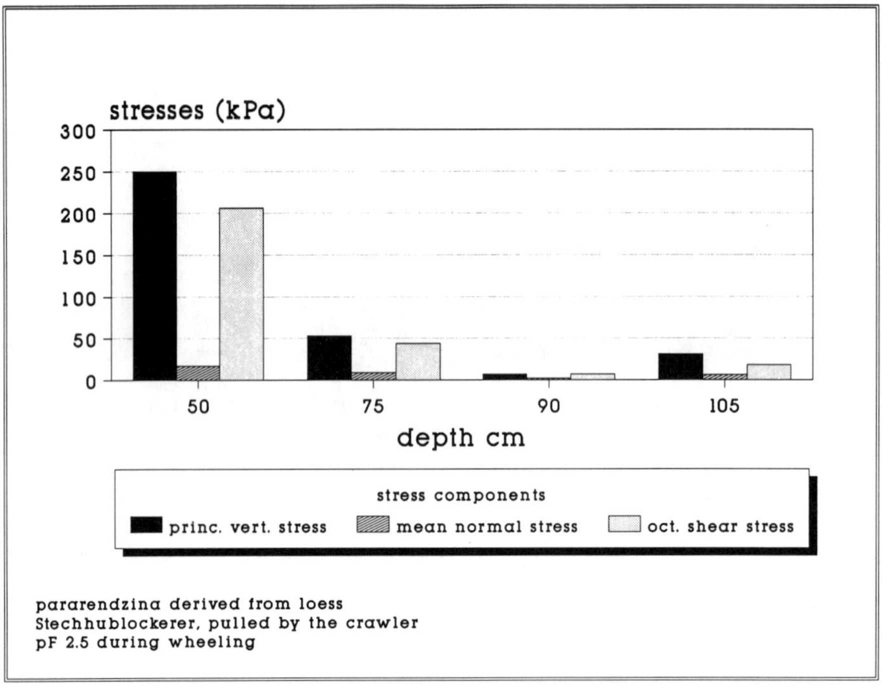

Figure 9. Stresses in between the reloosening tools (Stechhublockerer) and below during amelioration of an overconsolidated refilled soil (brown coal mining, Rheinbraun).

Jayawardane and Blackwell (1985) and Reich et al. (1985) developed a slotting technique and a ditch plow device, respectively, to segmentally reloosen soils. In such segmentally loosened soils the loosened area provide preferential pathways for water, air and root growth to depth. During trafficking, the undisturbed strong soil on either side of the trench should provide support for the wheels, thereby preventing recompaction of the weak loosened soil in the narrow trench or slot.

Figure 10 shows the soil reloosening technique by the ditch plow. The distances between the ditch are determined by the moldboard plow spacing

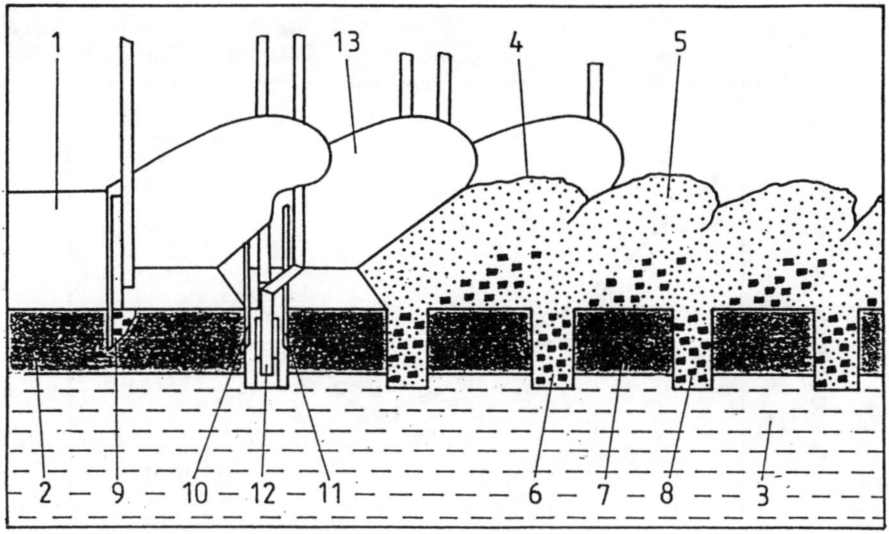

Figure 10. Schematic diagram of a ditch plowed soil. 1= Ap-horizon; 2 = compacted subsoil; 3 = undisturbed deeper subsoil; 4 = plowed, i.e., reloosened soil volume; 5 = basis of the plowed soil volume; 6 = ditch; 7 = unloosened soil material in between the ditchs; 8 = reloosened subsoil; 9-12 show special arrangements of the ditch plow technique. (From Werner et al., 1992.)

because the rigid tines are fixed exactly below the plows. The ditch plow device, which is mostly used in Germany, creates a reloosened soil volume of 30% up to 20 cm below the normal plowing depth, if the trench itself is 10 cm wide (for more detailed informations about the technique see Reich et al. 1985).

In slotted soils, the loosened soil volume varies with the slot configuration (Jayawardane et al., 1994). The stress distribution in the slotted soil volume mainly depends on the ratio between the slot width (Figure 11), the tire width as well as on the direction of wheeling (Blackwell et al., 1989). In slotted soils, both transverse wheeling (i.e., wheeling at right angles to the direction of the slot) and longitudinal wheeling (i.e., wheeling in the direction of the slot with the center of the travel path coincident with the center of the slot) led to a decrease in the vertical stresses in the slot. During transverse wheeling stresses only increased at slot widths greater than 220 mm. When the slots were wheeled longitudinally the slotted soil is less protected and higher vertical stresses were measured for any given slot width, when compared to transverse wheeling.

In undisturbed soil, the highest stresses were measured during the first wheeling. However, the formation of a more compacted soil volume near the soil surface during consecutive wheeling led to a more pronounced stress attenuation. The trend was reversed inside the slot. With repeated traversing, a recently slotted soil showed increasing stress values and became more compacted. Inside

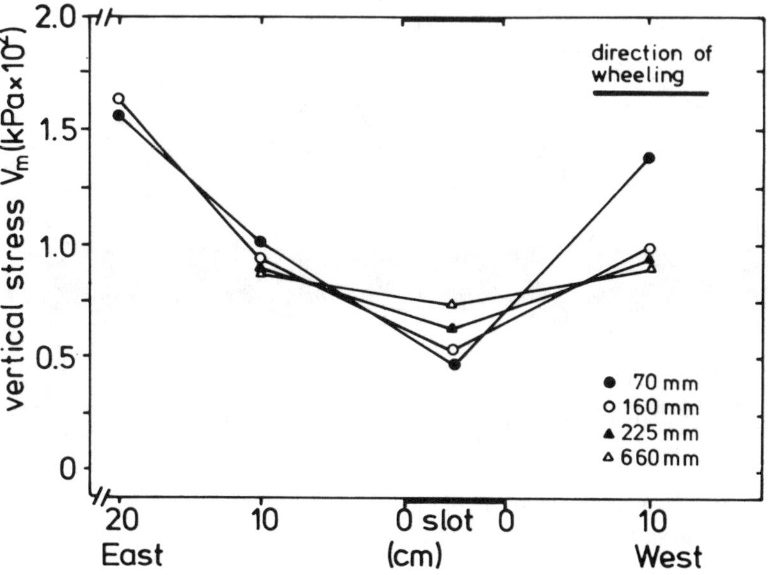

Figure 11. Vertical stress distribution inside and adjacent to the slot in dependence of the slot width.

a recently formed slot mean vertical stress values at all depths were similar regardless of the slot width. This could be due to reduced shear forces between the very loose slotted soil and the smoothed vertical slot wall. Thus any applied stress leads to compaction of the entire slotted soil volume.

Vertical stresses were also measured in ditch plowed gleyic luvisol derived from glacial till during trafficking with a tractor alone and a tractor pulling a slurry tank. The axle loads of the tractor and of the pulled slurry tank during these wheeling events were 8.3 kN and 29.3 kN respectively, while the contact area pressures were 0.13 MPa and 0.65 MPa. The water contents (Θv) of the soil were 0.23 in the top 10 cm and 0.18 (m^3 m^{-3}) at 20 to 50 cm depth. The corresponding pore water pressure values ranged between -24 and -11 kPa.

Repeated wheeling of the ditch plowed soil with 29.3 kN axle load led to higher stress values in the reloosened soil material in the beginning. After five wheeling passages, the stresses became more intensely attenuated in the center of the slot (Figure 12). During the first wheeling events additional soil material from the top layer was squeezed inside the trenched soil volume due to negative slip of the tire. This led to a significant bulk density increase. At 50 cm depth (i.e., below the trench) on the contrary, no differences in stress due to loading

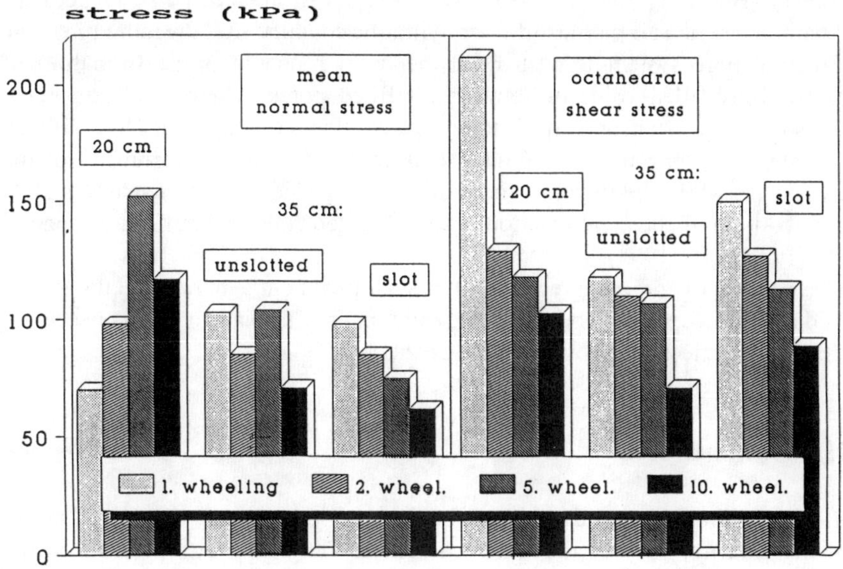

Figure 12. Stress distribution inside the ditch and adjacent undisturbed soil.

could be distinguished. The magnitude of negative slip effects can be derived from the octahedral shear stress values, which were much higher in the reloosened soil volume as compared to the undisturbed soil. Such slip processes will result in a very pronounced rearrangement of particles and thereby in recompaction, especially if consecutive wheeling is performed at too wet (i.e. weaker) soil conditions. Additionally, it has to be kept in mind, that shearing causes a rearrangement of particles and thus increase the access of water to soil particle surfaces. Thus, swelling induced strength reduction could occur (Horn, 1976). Furthermore, the mobility of soil particles is increased leading to further soil compaction during consecutive loading. While there is some protection of the macropore structure of the loosened soil in the slots and trenches, during trafficking of segmentally loosened soils, damage could be caused by excessive wheeling, especially in wet soil where the strength of the undisturbed soil is low.

VI. Effects of Compaction on Soil Properties

Irrespective of the amelioration technique used, the reloosened soil volume is more sensitive to further soil compaction during any further soil loading. Soil parameters like air permeability and saturated hydraulic conductivity are negatively affected as compared to both the natural soils and the "ameliorated" unloaded soils. In general, soil mixing always results in a complete disturbance of the pore system by destroying the existing aggregates and by increasing the tortuosity (Hartge and Horn, 1991). With respect to induced consequences for soil parameters like air permeability or hydraulic conductivity, the more intensive the natural pore system is destroyed, the more probable is the formation of isolated water filled voids in between air filled pores, which results in a 10^4 times smaller gas transport in those equivalent pores (Hartge and Horn, 1991). Soil water and ion transport are also reduced due to a higher tortuosity of the pore system, while the increased accessibility of particle surfaces may result in an altered type of chemical reactions and a changed filter and buffering capacity of the soil (Horn, 1994).

The process of soil reaggregation after such treatment generally requires long periods of time (up to several decades), and it mainly depends on the consecutive hydraulic and mechanical treatments of the soil.

VII. Conclusions

1. Soil reloosening and tractor wheeling in loosened soil require careful consideration not only of the water content and/or pore water pressure at the surface, but especially in the deeper soil horizons. As a rule of thumb, the pore water pressure at the loosening depth and in the deeper soil horizons has to be at least smaller than - 80 kPa in order to prevent the recompaction.

2. The reloosened soil volume has to be protected against recompaction during the following wheeling or tillage practices; thus, an extensive land use system and/or a reduced wheeling intensity has to be applied.

3. A complete soil reloosening has the advantage of a totally "improved" soil volume for the plant growth and improved permeability for gas and water. On the other hand, such soil volume is too unstable to carry the load of tractors, harvest devices, etc. It should also be pointed out that additional soil compaction during the loosening practices could occur due to lack of care in selecting the loosening technique (e.g., crawler mass, contact area etc).

4. In order to characterize or to classify the various loosening techniques it is necessary to measure not only physical soil parameters like bulk density, pore size distribution, and hydraulic conductivity, but it is mainly necessary to evaluate the stresses and the stress distribution inside the reloosened soil volume and the underlying material. Measurements are required of the three principal and three nonrectangular stresses at various positions in the soil during wheeling, in order to derive the principal stresses, the mean normal stresses, and octahedral shear stresses. The increased octahedral shear stresses at comparable mean normal stresses are mostly responsible for soil deformation and homogenization in both segmentally loosened and conventionally plowed soil.

5. Results from wheeling experiments on segmentally loosened soils show (e.g, for the ditch plow or slotting technique) the possible use of such technique to ameliorate overcompacted soils, with the advantage of a better preservation of the reloosened soil volume. However, the data in the literature also underline the very much increased sensitivity of such soil for recompaction under wet soil

conditions, even if they are only wheeled in rectangular direction to the trench, and with very wide tires. Soil deformation under wet soil conditions due to squeezing, shearing or kneading results in a poorer soil structure compared to the non-ameliorated soil, and should therefore be avoided.

VIII. Summary

Soil compaction leads to reduced water and air entry into surface and subsoils, associated problems of increased runoff and erosion, and consequently reduced root growth and crop yields. Increased compaction of agricultural lands in developed countries has been caused by a 3-4 fold increase in mass of agricultural tractors and the use of larger numbers (up to 14) of trafficking events in each cropping season. In recent years, the introduction of conservation farming practices has helped to reduce soil compaction. However, even with conservation farming, periodic loosening of the soil is required to maintain adequate flows of water and air into the subsoil. With heavily compacted soils, the quickest and most effective method of soil amelioration is mechanical soil loosening, followed by use of conservation farming practices and a reduced mechanical loading.

Management of soil compaction requires the understanding of the influence of soil physical properties and stress transmission patterns under trafficking, on the nature and extent of soil compaction. In order to prevent soil compaction, the mechanical strength of the soil as measured by the precompression stress, should exceed the vertical stresses developed under trafficking at all soil depths. An intensive deformation can occur due to high shear stresses, especially at high water contents and with increase in slip percentages.

The precompression stresses in soils vary widely with soil types and shows changes with depth, especially in soils with marked soil profile differentiation. The precompression stresses of soil layers are markedly reduced with soil loosening and with increase in soil water content, thus increasing the susceptibility of the soil to compaction. With soil loosening, the stress equipotential lines under trafficking are more vertically oriented compared to undisturbed soil, with an increase in the values of the concentration factor. This leads to deeper transmissions of stresses and hence an increase in the potential for compaction at depth.

Recompaction of loosened soils during subsequent trafficking can be minimized by limiting trafficking to times when soil is relatively dry, and by using traffic lanes. With use of crawler type tractors, compaction is reduced near the surface, but increased compaction can occur at depth. The other option is to segmentally loosen the soil, using slotting or ditch plows. The weak loosened soil in the slot or trench is largely protected from compressive stresses under wheeling by bridging of the wheel "contact patch" across the slot or trench, and the adjacent stronger undisturbed soil supporting much of the wheel load. However, damage could be caused by excessive wheeling, especially in wet soil, where the precompression stress of the undisturbed soil is low.

References

Baum, G. 1993. Bodenmechanische Untersuchungen zur Lockerungswürdigkeit von verdichteten rekultivierten Böden aus Löβ. Unpublished. Diplomarbeit Institute for Plant Nutrition and Soil Science, CAU, Kiel.

Blackwell, J., R. Horn, N.S. Jayawardane, R. White, and P.S. Blackwell. 1989. Vertical stress distribution under tractor wheeling in a partially deep loosened typic Paleustalf. *Soil Tillage Res.* 13:1-12.

Burger, N., M. Lebert, and R. Horn. 1988. Prediction of the compressibility of arable land. p. 153-157. In: J. Drescher, R. Horn, and M. de Boodt (eds.), Interaction of structured soils with water and external forces. *Catena Supplement* 11. ISBN 3-923381-11-5

Drescher, J., R. Horn, and M. de Boodt. 1988. Impact of water and external forces on soil structure. *Catena Supplement* 11. ISBN: 3-923381-11-5. 175 pp.

Dumbeck, H. 1991. Bodenphysikalische Untersuchungen an wiederaufgelockerten Pararendzinen aus Löß. *Mitt. Dtsch. Bodenkde. Ges.* 66:161-164

Eck, H.V. and P.W. Unger. 1985. Soil profile modification for increasing crop production. *Adv. Soil Sci.* 1:65-100.

FBW. 1989. 3. Waldschadensbericht der Bundesregierung. *Verlag KFK Karlsruhe.* 400 pp.

Grath, T. 1993. Effects of soil compaction on physical, chemical and biological soil properties and crop production. Swedish University of Agricultural Sciences, No. 5, ISSN 0348-0976.

Hakansson, I., W. Voorhees, and H. Riley. 1988. Weather and other environmental factors influencing crop responses to tillage and traffic. *Soil Tillage Res.* 11:239 - 282.

Hartge, K.H. and R. Horn 1991. *Einführung in die Bodenphysik.* 2nd. Ed. Enke Verlag.

Hempfling, R., H.R. Schulten, and R. Horn. 1990. Relevance of humus composition for the physical/mechanical stability of agricultural soils – A study by direct pyrolysis-mass spectometry. *J. Anal. Appl. Pyrolysis* 17:275-281.

Horn, R. 1976. Festigkeitsänderungen infolge von Aggregierungsprozessen eines mesozoischen Tones. PhD Dissertation. Hannover, Germany. 76 pp.

Horn, R. 1981. Die Bedeutung der Aggregierung von Böden für die mechanische Belastbarkeit in dem für Tritt relevanten Auflastbereich und deren Auswirkungen auf physikalische Kenngrößen. Schriftenreihe FB 14, TU Berlin, ISSN:379830792.

Horn, R. 1986. Auswichung unteraschiedlicher Bodenbearbeitung auf die mechanische Belastbalkeit von Ackerböden. *Z. Pflanzenerz und Bodenkunde* 149:9-18.

Horn, R. 1993. Mechanical properties of structured unsaturated soils. Special Issue *Soil Technology* 6:47-75.

Horn, R. 1994. The effect of aggregation of soils on water, gas and heat transport. p. 335-361. In: E.D. Schulze (ed.), *Fux Control in Biological Systems.* Academic Press.

Horn, R. and M. Lebert. 1992. Beurteilung der Meliorationsfähigkeit von Waldböden anhand von bodenphysikalisch/-mechanischen Kenngrößen. *Allgemeine Forstzeitung* 19:998-1005

Horn, R., N. Burger, M. Lebert, and G. Badewitz. 1987. Druckfortpflanzung in Böden unter fahrenden Traktoren. *Z.f. Kulturtechnik und Flurbereinigung* 28:94 - 102.

Horn, R.L., C. Johnson, H. Semmel, R. Schafer, and M. Lebert 1992. Stress measurements in undisturbed unsaturated soils with a stress state transducer (SST)-theory and first results. *Z. Pflanzenerz und Bodenkunde.* 155:269 - 274

Horn, R. and A.R. Dexter. 1994. Effect of wheeling speed on slip affected stresses in soil. *Soil Tillage Res.* (in preparation)

Horn, R. H. Kretschmer, H. Semmel, K. Bohne, T. Baumgartl, A. Neupert, and U. Zöllner. 1994a. Soil mechanical properties of a partly reloosened (ditch plow system) and a conventionally tilled overconsolidated gleyic luvisol derived from glacial till. *Soil Tillage Res.* (submitted).

Horn, R., T. Baumgartl, S. Kühner, W. Graesle, R. Raper, T. Way, and C. Johnson. 1994b. Volumetric stress-strain distribution in structured unsaturated soils during wheeling – theory and measurements. *J. Agric. Engineer.* (submitted).

ISTRO. 1991. Soil Tillage and Agricultural Sustainability. 12th Intern. Conf. Ibadan, Nigeria. 682 pp.

Jayawardane, N.S. and J. Blackwell. 1985. The effect of gypsum enriched slots on moisture movement and aeration in an irrigated swelling clay soil. *Aust. J. Soil Res.* 23:481-492.

Jayawardane, N., J. Blackwell, G. Kirchhof and W.A. Muirhead. 1994. Slotting – a deep tillage technique for ameliorating sodic, acid and other degraded subsoils and for land treatment of waster. p. 109-146. In: N.S. Jayawardane and B.A. Stewart (eds.), *Subsoil Management Techniques.* Adv. Soil Sci., CRC Press, Boca Raton, FL.

Larson, W.E., G. Blake, R. Allmaras, W. Voorhees, and S. Gupta. 1989. Mechanics and related processes in structured agricultural soils. NATO ASI Series, E: Applied Sciences 172. Kuwer Academic Publishers. 180 pp.

Lebert, M. 1991. Stabilitätseigenschaften rekultivierter Böden aus Löß im Rheinischen Braunkohlen-Tagebaurevier.Mitt.Dtsche. *Bodenkde. Ges.* 66:165-168

Nichols, T.A., A.C. Bailey, C.E. Johnson and D. Grisso. 1987. A stress state transducer for soil. *Transactions ASAE*: 30(5):1237-1241.

Reich, J., H. Unger, H. Streitenberger, C. Mäusezahl, C. Nussbaum, and P. Steinert. 1985. Verfahren und Vorrichtung zur Verbesserung verdichteter Unterböden. EB DDR Nr. 233915, described also in Reich, J., H. Streitenberger, and N. Romanenko. 1991. *Agratechnik Berlin* 41:57-62.

Scheffer, F., and P. Schachtschabel. 1992. Lehrbuch der Bodenkunde 13th edition, *Encke Verlag,* Stuttgart ISBN: 3-43284-773-4.

Schulte Karring, H. 1988. 150 Jahre Technik der Tieflockerung. Eigenverlag Landes-, Lehr und Versuchsanstalt für Landwirtschaft, Weinbau und Gartenbau. 5483 Bad Neuenahr Ahrweiler Germany.

Schulte Karring, H. 1990. Soil physical properties of the Abbruchlockerer during soil amelioration. Proc. 14 th. Int. Soil Sci. Soc. Conf. Kyoto, Japan I: 342-344

Schulte Karring, H. 1991. Die Mehrzweck-Abbruch-Tieflockerung mit Düngung, ein neues Verfahren zur Melioration von verdichteten oder kontaminierten Böden. *Z. Kulturtechnik und Landentwicklung* 32:48-59.

Semmel, H. 1993. Ph.D. Thesis. Auswirkungen kontrollierter Bodenbelastungen auf das Druckfortpflanzungsverhalten und physikalisch-mechanische Kenngrößen von Ackerböden. Schriftenreihe Inst. Pflanzenern. u. Bondnkde. 196S.

Soane, B.D. and C. van Ouverkerk (eds.). 1994. *Soil Compaction.* Elsevier. (in press).

Werner, D., D. Roth, J. Reich, C. Mäusezahl, U. Pittelkow, and P. Steinert. 1992. Verfahren der Unterbodengefügemelioration mit dem Schachtpflug B 206 A. Internal report of the Landwirtschaftliche Untersuchungs – und Forschungsanstalt Thüringen. 92 pp.

Identification and Manipulation of Soil Biopores for the Management of Subsoil Problems

K.E. Lee and K.R.J. Smettem

I. Introduction

The impetus for this review stems from the growing recognition that effective management of soil burrowing organisms may provide a sustainable, low cost, practical approach for ameliorating subsoil problems. Several recent reviews are also pertinent to this topic. From a physical perspective, the reviews by Beven and Germann (1982), White (1985) and Bouma (1994) give comprehensive coverage. Curry and Good (1992) give a broad review of soil faunal degradation and restoration, Syers and Springett (1984) describe the potential benefits to soil fertility that might accrue through encouraging earthworm activity and Kladivko and Timmenga (1990) review management influences on earthworms.

Aspects of soil structure that affect topsoil, the habitat of a large and varied biomass, were distinguished by Hamblin (1985, 1991) from those that affect the subsoil. In general, the subsoil is considered to function as a storage for water and plant nutrient reserves and includes a comparatively small biomass. Hamblin

1-56670-020-5/95/$0.00+.50

(1991) proposed that in the topsoil aggregate stability and infiltration capacity are of prime importance, whereas in the subsoil the critical structural condition is the extent of vertically continuous macroporosity relative to the total porosity, *i.e.*, the balance between water storage and transmission properties. In the context of water infiltration and aeration, macropore continuity between the topsoil and the subsoil is also of particular importance.

This chapter reviews how the soil environment may be modified by the burrowing activity of soil animals. Specifically, we are concerned with the nature, extent and role of soil biopores, a class of soil macropores of varying diameters and depths that are created by soil organisms. We do not include cracks or macropores that are left when roots die and decay, but only those that are formed by soil animals.

In order to understand the contribution that biopores may make to the amelioration of subsoil problems it is necessary to review firstly the excavation and construction of biopores, together with their biological significance for the organisms that construct them. Information is then required on the dimensions and extent of soil biopores, their morphology and their influence on soil physical properties. With this knowledge, consideration can then be given to the dispersal of small soil organisms and the growth of plant roots through biopores and their access to water and nutrients, particularly in the deeper soil layers. Finally, we discuss possibilities for the management of soils and soil organisms to enhance the beneficial effects of burrowing activity.

II. Types of Pores Formed by Soil Animals

Ringrose-Voase (1991) distinguished two major categories of pore space in soils: structural pore space, the spaces in soils that define structural units (peds and aggregates), and non-structural pore space, spaces that are not related to structural units, but may be within structural units or may traverse several structural units. He distinguished some major types of non-structural pores, including channels, the more or less cylindrical macropores that mark the pathways of roots and the burrows of soil fauna, and chambers, including the cavities excavated as resting places at the ends of burrows of earthworms or other solitary soil animals, and galleries, nests and storage spaces, excavated by soil-inhabiting social animals, the ants and termites.

Hole (1981) distinguished voids formed as enclosures from those formed by removal. Among soil animals that make removal voids he recognized fossers, equipped with digging limbs, miners, such as ants and termites, that seize single grains or aggregates with their mandibles and carry them away, and tunnelers, such as earthworms, that compress or swallow soil, which is then excreted as casts at the surface or in existing voids.

A. Earthworm Burrows

Earthworms are found in the soils of all but the coldest and the driest regions of the earth. Their burrows are the most numerous and most important macropores of animal origin in nearly all soils (see Lee 1985; Lee and Foster 1991). As for all soil animals, there is much within-site variation in earthworm population density, related to within-site variation in soil properties, but populations and their burrows are more randomly distributed (Edwards and Lofty, 1977) than those of ants and termites.

1. Ecological Groups of Earthworms and their Channels

Three major ecological groups of earthworms have been recognized: epigeic species, that live and feed above the surface of the mineral soil and do not make burrows; anecic species, that feed at the surface and live in permanent burrows that open at the soil surface and endogeic species, that live and feed below the soil surface, making extensive burrow systems and ingesting and casting much soil as they seek soil organic matter for food (Lee 1959, Bouché 1977, Lee and Foster 1991). Epigeic species are common only in forests and similar plant associations, where the accumulation of surface litter provides a habitat for them. Anecic and endogeic species are also found in forest soils, but they are the common earthworms of agricultural lands, and have an important influence on land use practices.

The burrows of anecic species are predominantly vertical, and may penetrate to depths of a meter or more (see Lee and Foster 1991). They are the more or less permanent "homes" of individual earthworms, and are maintained for the life of the resident earthworms. These burrows are commonly regarded as conduits into the deeper soil horizons for ponded water, but structural features of the burrows and behavior of anecic species that prevent water from easily entering their burrows negate this simplistic view.

Burrows of many anecic species have multiple entrances, and side branches, sometimes leading to chambers in which cocoons are deposited or where litter brought from the surface is stored. Kobel-Lamparski and Lamparski (1987) described such burrows, made by *Lumbricus badensis* and penetrating more or less vertically to a depth of about 2 m. Near the bottom of the burrow they recorded that the humus-stained lining became discontinuous and then virtually absent, with the burrow terminating in a more or less horizontal, branched section. This form of burrow must give the resident earthworm advantages and also affect the rapid drainage of ponded surface water down the burrow. If the earthworm has no alternative but to stay in a simple burrow with one entrance, water ponded at the surface would run into and fill the burrow above its body. Oxygen in this trapped water would quickly be used up by microorganisms, forcing the earthworm to abandon the burrow or asphyxiate; the availability of alternative branches of the burrow opening to the surface would provide the opportunity for the earthworm to lie in one branch close to the soil surface while water ran freely down other branches. The lack of burrow wall lining and horizontal orientation of the burrow at depth must enhance the infiltration of water into deep soil layers. In these circumstances it would be reasonable to regard the burrow as a simple conduit for water to drain quickly into the subsoil.

The entrances of surface-opening burrows may be partly or completely sealed by the resident earthworms, by plugging them with leaves collected from the surface or with soil particles or aggregates. For example, burrows of *Aporrectodea caliginosa* and *A. rosea* that penetrated surface soil crusts were shown by Roth and Joschko (1991) to be quickly sealed by the earthworms after the onset of simulated rain. With rain at a rate of 30 mm h^{-1} burrows <4-5 mm diameter were nearly always sealed within 15 minutes of the onset of rain, while burrows >5 mm diameter remained unsealed for about 45 minutes. This temporal variation in sealing meant that the smaller diameter burrows permitted little infiltration, while the larger burrows markedly increased infiltration.

The burrows of endogeic species have predominantly horizontal orientations. Their casts are mainly deposited within the burrows, so that voids derived from their burrows are found in the soil as generally elongated in form, of irregular lengths and with diameters determined by those of the earthworms that made them. Some of these voids are isolated by plugs formed by casts, while some are connected by more or less randomly positioned, narrower voids around casts that partially block the burrows.

Endogeic earthworms are geophagous, i.e., they ingest soil, although examination of the gut content of some species has shown that they selectively feed on organic materials in the soil. Their diet is low in energy compared with that of anecic species, which feed at the surface on (mainly dead) plant tissue, so they must ingest large quantities of soil to satisfy their metabolic needs. Lavelle (1983) showed that *Millsonia anomala*, an endogeic species from Ivory Coast savanna lands, consumes soil at a rate of 20-36 times its body weight per day, so that a field population of 215,000 per hectare would ingest about 500 t of soil per hectare per year; when the consumption of *M. anomala* is added to that of two other geophagous species that are found at the same site, the three species together ingest 850-1200 t of soil per hectare per year. The endogeic species *Aporrectodea rosea,* in an English beech wood consumes soil at a rate of about 2-4 times its body weight per day, so that a field population of about 340,000 per hectare would ingest about 10-15 t of soil per hectare per year (see Bolton and Phillipson,1976; Phillipson and Bolton, 1977). Reasons for the great difference in consumption rates relate to the lower organic matter content of the Ivory Coast soil compared with the European soil and to the higher metabolic rates of tropical compared with temperate earthworm species. The biological significance of these two sets of data is further discussed by Lee (1985). The combination of burrows of anecic and endogeic species provides pathways for ponded water to infiltrate into the soil and for its vertical and lateral movement into and through the subsoil, but it is important to recognize that these macropores are not necessarily continuous over their full length, or of uniform diameter throughout their length, although Kretzschmar (1987a) and Kretzschmar and Monestiez (1987) considered that water and gas flow through them is more or less unimpeded, comparable to those at the soil surface. Kretzschmar (1987a) examined the significance of burrows as pathways for diffusion processes from the surface into the soil, using simulated burrow systems. It was shown that the burrows lead to a partitioning of soil space and the shortening of diffusion pathways.

2. Mechanics of Burrow Construction

Earthworms are tunnelers that use two basic mechanisms to make burrows. These are: 1. exerting axial pressure to insert the anterior portion of the body between soil aggregates, then compressing the soil radially using cyclic contraction and relaxation of the muscles to generate peristaltic waves of pressure along the body; 2. everting the pharynx and using suction to engulf and swallow soil, which is passed through the gut and excreted as molded casts. A combination of these two mechanisms is probably used most often.

Trueman (1978) measured internal celomic pressures of *Lumbricus terrestris* at rest and found little fluctuation from a value of <200 Pa. During locomotion a maximum value of about 7500 Pa has been measured (Seymour, 1978). For a non-rigid body wall the radial force exerted during burrow construction would be equal to the internal pressure times the area of contact. Seymour (1978) however, demonstrated that *Lumbricus terrestris* was able to exert an "excess lifting force" of about 145% and attributed this to the intrinsic rigidity that can be developed by contracting and stiffening body wall muscles.

The axial and radial pressures exerted by *Aporrectodea rosea* were measured by McKenzie and Dexter (1988a, 1988b). The earthworms were burrowing through discs about 10 mm thick, moulded from a sandy loam, and compressed to provide a range of strengths. Measured axial pressures exerted by the worms were in the order of 100 kPa and radial pressures were shown to exceed 230 kPa. Calculation of axial pressures assumed that the earthworm exerted pressure evenly across the cross-sectional area of the worm, i.e., the cross-sectional area of the burrow. It is likely that the pressure is exerted over a smaller anterior area of the earthworm, i.e., the prostomium and peristomial region. McKenzie and Dexter (1988a) calculated that, for *A. rosea,* if all the pressure were exerted over the frontal area of the prostomium alone, the maximum axial pressure exerted might be as high as 1 MPa, but they concluded that such a high pressure was unlikely, because it could damage the prostomium. It seems most likely that pressures exerted by earthworms are at the lower end of the range of those exerted by plant roots. Some burrowing through compact layers may result from the ability to elongate and narrow the anterior segments, to penetrate into small cracks, followed by the exertion of radial pressure to enlarge a crack so that the full diameter of the body can be accommodated.

More generally, the penetration of compact layers must depend on the habit of earthworms of everting the pharynx to cover an area of soil, which is moistened with saliva and then removed by suction and swallowed. The low pressures exerted would also explain the observation of Dexter (1978) that the soil is not compacted or cracked around earthworm burrows in compact soil horizons, even though the rate of burrowing of *Aporrectodea caliginosa* in small cores was shown not to be affected at soil strengths up to penetrometer resistances of 3 MPa.

3. Dimensions, Numbers and Distribution of Earthworm Burrows

Earthworm burrows are circular or near-circular in cross section and range in diameter from about 1 mm to >10 mm, depending on the size of the earthworm constructing the burrow. Because the size of earthworms varies within and between species they produce a range of burrow diameters.

Barley (1959) presented data on burrow sizes in a loamy permanent pasture near Adelaide, South Australia. This site supported the lumbricids *Aporrectodea rosea* and *A. caliginosa* and tunnel diameters ranged from 0.5 to 3.5 mm with a mode of 1.1 mm and an arithmetic mean of 1.4 mm. For an arable loamy sand in Southern England, Omoti and Wild (1979) reported a biopore size distribution of 2.0 to 10.0 mm diameter with a modal size class of 2.0 to 5.0 mm. This modal class is the same as reported by Ehlers (1975) for a loess soil in Germany. In both these soils the larger lumbricids *Lumbricus terrestris* and *Aporrectodea longa* are known to be present. Some species construct larger burrows, but these are generally restricted to deeper soil layers under forest or sometimes in grasslands (see Lee, 1985) and so are of little interest in an agricultural context.

The length of an individual burrow is again dependent upon the ecological group of the species constructing it, its function and soil conditions. Barley (1959) observed that the tunnels of *A. rosea* and *A. caliginosa* were frequently irregular in the top 15 cm but below this depth they seldom branched and extended at a small angle to the vertical to a depth of about 0.5m. The lower ends of the tunnels were frequently enlarged to form an ovoid chamber, indicating that they were constructed by worms entering diapause. *L. terrestris* and *A. longa* are known to burrow much deeper into the soil. *L. terrestris* is unable to resist drought by either quiescence or diapause (Bouche and Gardiner, 1984) and burrows must therefore extend to depths that ensure the animal can reach zones of adequate moisture supply during dry periods. Depths of 1.5 m are not uncommon and Lee (1985) cites reports from Ohio, U.S.A. of *L. terrestris* burrows to 3 m. Kobel-Lamparski and Lamparski (1987) point out that although *L. terrestris* is normally a deep digging species it can be confined to the surface layers in forests. Similarly, *L. rubellus* can be exclusively near-surface dwelling in forests with a litter layer (Wilke, 1953), while in open areas without a litter layer even small juveniles switch to deep digging (Kobel-Lamparski and Lamparski, 1983). The same authors also demonstrated that, during its development, *L. badensis* also changed from surface burrowing to deep digging. There are many reports of numbers of earthworm burrows ranging from 100 to 300 per square meter of soil surface area, and up to 2,000 per square meter have been counted (see Lee, 1985; Lee and Foster, 1991). The deepest burrow reported for a lumbricid was that of a giant species (*Scherotheca* sp) at 5.20 m (Bouche, 1971); the lumbricids that inhabit much of the pasture and cropland soils of temperate regions of the world live in semi-permanent burrow systems that sometimes penetrate to depths of 1 m, but most commonly do not extend below 0.2 - 0.5 m depths.

Barley (1959) estimated the number of earthworm burrows and root channels at various depths in a 10-year-old pasture on an Alfisol in South Australia. Numbers of earthworm burrows recorded were 1000 m^{-2} at 0.05 m depth, 300 m^{-2} at 0.2 m and 75 m^{-2} at 0.5 m. After four years of no-tillage practice on a soil formed from loess, Ehlers (1975) recorded increases in number of earthworm burrows from 27 to 117 m^{-2} at 0.02 m depth, from 79 to 141 m^{-2} at 0.2 m, from 187 to 305 m^{-2} at 0.30 m, and from 348 to 363 m^{-2} at 0.6 m.

Kretzschmar (1978, 1982) measured the lengths and volumes of earthworm burrows, which he took to be burrows 2 mm or more in diameter, in a pasture soil at Citeaux, France. To a depth of 1.2 m, burrows varied seasonally in length from 142 m m^{-2} to 888 m m^{-2} over a period of 1.5 years. The corresponding burrow volume, which must represent most of the macropores with equivalent cylindrical radius >1 mm diameter, varied seasonally from 1.3 m^3 to 9.2 m^3. Zicsi (1983) estimated that the European endogeic lumbricid *Lumbricus polyphemus* consumed soil (dry weight) at a rate of 0.1 times its body weight per day, so that a field population he estimated as 40,000 ha^{-1} would ingest about 8.1 t of soil per hectare per year. Zicsi estimated, from experiments with earthworms kept in monoliths of undisturbed soil, that about 25% of the soil ingested by *L. polyphemus* was cast at the soil surface. Assuming a bulk density of the soil of about 1.4 t m^{-3}, this represents a volume of macropores of about 0.6 m^3.

4. Rate of Burrow Development

Barley (1959) reported that specimens of *A.caliginosa* of weight 0.2 g consume 0.3 times their own weight of oven dry earth per day at 15° C. He assumed that 0.2 g was a mean weight; hence from his data on pore sizes we may assume that this corresponds to a mean pore size of 1.4 mm. Given a bulk density of 1.4 Mg m^{-3} the corresponding rate of excavation would be 2.8 mm per day. This rate is considerably less than reported by Bolton and Phillipson (1976) for *A. rosea*. In repacked laboratory chambers they measured rates of up to 100 mm per day immediately following introduction of an earthworm to the chamber. After 2-3 days a more constant level of about half this value was attained. These rates were considered to result from a laboratory induced escape reaction. However after day 4 the worms were relatively inactive and burrowed only for the purpose of feeding. At this time, rates of 40 mm per day were still observed for adults and 10-25 mm per day for small immature worms.

5. Burrow Linings

Bouché (1975) defined a zone 2 mm in thickness around earthworm burrows and named it the drilosphere. Bhatnagar (1975) demonstrated that in a French pasture soil the drilosphere, which made up no more than about 2% of the total soil volume, contained about 40% of the total soil population of aerobic nitrogen-fixing organisms, 13% of the anaerobic nitrogen-fixers and 16% of the denitrifiers, with maximum concentrations at 20-40 cm depth in the soil. Bhatnagar (1975) attributed the preponderance of aerobic nitrogen-fixers in the drilosphere to the enrichment of the drilosphere with decomposing organic matter deposited there by earthworms; he also considered that the generally low C:N ratios of soils with high earthworm populations result from the stimulation of aerobic nitrogen fixers and consequent net increase in nitrogen in the drilosphere. The nitrogen metabolism of earthworms is discussed in Lee (1983).

Lavelle et al. (1989) have further pointed out that the nitrogen-rich mucus secreted by epithelial cells, to maintain a moist surface for respiratory exchange and provide a lubricant for earthworms as they move through their burrows, introduces readily available nitrogen into the drilosphere and so stimulates microbial and plant metabolism. The quantities of mucus secreted have not been much studied, though Needham (1957) estimated that about 50% of all nitrogen excreted by *Lumbricus terrestris* and *Eisenia fetida* was included in components of mucus. Krishnamoorthy (1985) estimated that 600-800 kg ha^{-1} y^{-1} (dry weight) of cutaneous mucus was secreted by earthworms in grassland and woodland sites in southern India. Nitrogen is also excreted as ammonia and urea, through segmentally arranged nephridiopores, and this nitrogen must also be added to the drilosphere. Lavelle et al. (1989) likened the drilosphere to the rhizosphere, with its similar production of exudates, energetic equivalents of intestinal and cutaneous earthworm mucus, the selective activation of soil microflora and also modifications of soil structure. Inputs of nutrients to the drilosphere must be more concentrated around the burrows of anecic earthworms that stay in a "home" burrow than around those of endogeic earthworms that continuously extend their burrows and move on.

Loquet (1978) measured carbon dioxide output and invertase, urease and dehydrogenase activity in the drilosphere; there were fluctuations throughout the year, with activity sometimes greater, sometimes less than in surrounding soil, but with no regular pattern. Ross and Cairns (1982) grew rye grass on subsoil in the presence or absence of *Aporrectodea caliginosa* and found that activities of invertase, amylase and phosphatase, and plant growth were stimulated by the presence of the earthworms; they did not claim to have proved a causal relationship between the presence of earthworms and increased enzyme activity, concluding that it was more likely that the increased plant growth, and therefore in plant material available for decomposition, probably stimulated enzyme activity.

a. Anecic Species

Burrows of anecic species characteristically have a distinct lining that includes decomposed plant tissue, apparently derived from casts deposited in the burrow and from remnants of plant leaves pulled into the burrow to be used as food or to block the burrow entrance, together with coatings deposited from solution in the well aerated burrow, coated with mucus secreted from the epidermis, with the surface smoothed and mixed with mineral soil materials by the passage of the earthworm's body.

Jeanson (1964, 1971) showed that such burrows, made by *Lumbricus terrestris* and *Aporrectodea icterica* in laboratory cultures, had a continuous lining, about one tenth of the burrow diameter, with a surface layer of oriented clay particles overlying three concentric layers, successively of humic material, calcium carbonate and iron oxides. After eight months of occupation, electron probe analysis showed that burrows of *L. terrestris* had concentrations of iron, manganese and calcium in the lining layers, with the iron and manganese evenly distributed and declining progressively in concentration so that they were indistinguishable at about 600 μm from the surface, while the calcium was concentrated in rounded "corpuscles" of calcite, a few micrometers in diameter. Jeanson (1961) had concluded that the superficial, dark colored layer consisted of excreted casts, spread on the surface by the movement of the earthworms. The iron and manganese concentrations probably reflect an increase in redox potentials at and near the boundary of air-filled burrows, resulting in oxidation and deposition of oxides. The "corpuscles" of calcium carbonate may be excretions from the calciferous glands that are included in casts, similar to but of smaller size than those reported by Wiececk and Messenger (1972) in the anterior portion of the alimentary canal of *L. terrestris* and in soils where they lived. A more comprehensive discussion of redox potentials as they affect earthworms, and of the products of calciferous glands as they affect soils, is given by Lee (1985).

b. Endogeic Species

There have not been detailed studies of the structure of the walls of burrows made by endogeic species, as there have been for those made by anecic species. Kretzschmar (1987b) noted a biogenic (cast) layer about 1 mm thick on the walls of burrows of the endogeic species *Aporrectodea longa* and *A. caliginosa,* coated with a superficial layer, 30-80 μm thick, apparently formed by layers of mucus, each 9-10 μm thick, deposited each time the earthworms move through their burrows. The accumulation of other layers, such as oxides of iron and manganese, must depend on the longevity of the burrows after the earthworms have abandoned them. Lee (1959) has observed recognizable burrows of the New Zealand endogeic species *Octochaetus multiporus,* heavily coated with humic and ferric-iron oxide layers, in the B horizons of palaeosols formed from loess during Pleistocene interglacial periods. Similar earthworm-burrow-like channels, which may be of great age, are common in many other circumstances, e.g., in the B horizon of "giant" podzols in coastal regions of Queensland.

B. Burrows and Nests of Termites and Ants

Termites and ants are social animals that live in concentrated nests, with nest populations that range from hundreds to millions of individuals. There are some termites that construct their nests entirely above ground, in trees, logs, or timber, without using soil materials. Most termites, however, use soil or mixtures of soil and organic materials (excreta, saliva) for construction. The burrows and chambers that they excavate are the nests of subterranean species, food storage spaces, chambers from which soil materials are obtained for the construction of above-ground nests, and burrows that provide access to and pathways between the nests and other chambers, and access to food and water. The excavations and constructions associated with a termite nest are part of the more or less sealed living and feeding space centered on the nest with its colony of termites. Lobry de Bruyn and Conacher (1990) summarized data on the soil surface area directly affected by termite nests and mounds; estimates generally ranged between 0.3% and 3% of the ground surface, but in exceptional circumstances up to 20% was affected.

1. Ecological Groups of Termites and their Burrows

Lee and Foster (1991) recognized three major functional groups of termites, as follows.

- Species that build nests above the ground surface, in trees and logs, that excavate and use some soil, often mixed with excreta, for nest construction.
- Species that build mound nests on the ground surface, sometimes extending into excavated galleries below the ground.
- Species that excavate their nests below the ground surface.

2. Mechanics of Burrow and Nest Construction

Termites and ants are principally miners, in the terms used by Hole (1981), but some species have been observed to make burrows by compressing soil outwards, with the head or the whole body, or to ingest and regurgitate soil, while some, especially the "humus feeders" ingest soil and pass it through the gut as do earthworms (see Lee and Foster 1991). Many species simply excavate soil and dump it on the surface when subterranean nests, tunnels, or chambers for the storage of food are constructed, while others use the soil they excavate to build above-ground nests or mounds.

3. Distribution and Depth of Termite Burrows

Mound construction is accompanied by the excavation of access tunnels into the subsoil and of chambers for storage of food or fungus combs, or from which the termites mine soil for construction. The burrows commonly cover a much wider area than that of the nest itself. Unlike the burrows of endogeic earthworms, which are constantly extended in search of rather diffuse food supplies, the burrows associated with termite nests are built as "highways" that connect the inner nest space with the ground surface or provide access to substantial sources of food, water, etc. They are maintained, enlarged and used as necessary for the life of the colony, which is commonly up to ten years and for some species may be much longer, or until a particular food source is exhausted. Ratcliffe and Greaves (1940) estimated that a mound colony of *Coptotermes lacteus* had burrows leading to food sources that extended over an area of 0.6 ha, with individual burrows radiating up to 50 m from the mound. A mound of *Macrotermes michaelseni* in Kenya was shown by Darlington (1982) to have associated with it 6 km of subterranean galleries and 72,000 subterranean storage chambers, extending over an area of 0.8 ha. Several other entirely subterranean termite species were also present, so the total length of termite burrows could easily have been twice that attributed to *M. michaelseni.*

Termites that live in arid to semi-arid regions excavate vertical tunnels to provide access to the water table, and transport water, which is swallowed, carried to the nest and then regurgitated. Such tunnels commonly penetrate more than a meter, and there are records of galleries at depths as great as 70 m (see Lee and Wood, 1971b).

4. Particle Size and Origin of Soil Removed from the Profile During Construction of Burrows and Mounds

The particle or aggregate size of soil that is simply removed and dumped at the surface is limited only by the sizes that can be grasped and carried by the insect. Nye (1955) compared the particle sizes of soil particles derived from a red earth near Ibadan, Nigeria. The soil is used by termites to form thin soil sheets over fallen twigs on the ground surface, piled loosely on the soil surface by a species of small ants and by a species of large ants, and is also deposited as casts by casts of the earthworm *Hippopera nigeriae*. The data are shown in Table 1 and clearly illustrate the differences between size fractions that are preferentially selected by tunnelers, animals that swallow soil and pass it through their gut (in

Table 1. The size distribution of material deposited on the surface of a red earth by different soil fauna

Animal group	Percent material		
	>0.2 mm	>1.0 mm	>2.0 mm
H. nigeriae	16	0	0
Termites	40	4	0
Small ants	35	10	0
Large ants	77	27	5

(From Nye, 1955.)

this case, only *H. nigeriae*), and miners, animals that grasp and carry soil, without swallowing it (the termites and ants). *H. nigeriae* is a large animal compared with termites and ants, but its casts contained only 16% of material of particle size >0.2 mm and none >1.0 mm.

Excavated soil that is used by termites in the construction of above-ground nests, runways, soil sheetings, etc., is selected particularly for its moisture and its clay content, i.e., its suitability for molding into the fabric constructed by the termite. Lee and Wood (1971a, 1971b) used clay mineralogy to trace the origin in soil profiles of materials used for above-ground construction by a variety of Australian termites. They found that, within rather broad limits, clay was selected preferentially compared with coarser soil particles, and that the soil was most commonly excavated from the B horizon of soil profiles, especially where the A horizon has a low clay content. Even in a subterranean nest constructed by *Drepanotermes rubriceps* there was evidence of clay enrichment by removal to the surface of sand, and the flat nest surface, level with the ground surface, was covered by a thin layer of sand, apparently that removed from the soil during nest construction.

5. Burrow Linings

Termite burrows, especially in semi-arid regions, become semi-permanent features of soil profiles; they include burrows that are predominantly vertical and predominantly horizontal. Like earthworm burrows, they become lined with iron oxide, precipitated from the soil solution in the aerobic environment within the burrows. Water and gas flow through them must be more or less unimpeded, so that they have the effect of partitioning the soil space and shortening diffusion pathways. It has been proposed by a number of authors beginning with Nazaroff (1931) that extensive areas of vesicular laterite and laterite crusts result from the deposition of iron oxides in old termite nests and galleries. The theory derives from uncritical comparisons of superficial resemblances between the laterites and termite constructions (see Lee and Wood, 1971b, for more information).

Construction and maintenance of these tunnels and chambers are essential to the social organization of termite colonies and to the survival of the individual termites. They are lined by the termites either with excreta or soil, laid down in laminae, sometimes of excreta or soil alone, more commonly in alternate layers of excreta and soil, or with flattened pellets of excreta or soil or mixtures of the two, plastered onto the burrow walls and stuck together with a layer of excreta or saliva. The micro-fabric of burrow walls and nest materials of 14 species of Australian termites were described by Sleeman and Brewer (1972). The excreta and excreta-soil mixtures that line many termite burrows closely resemble carton, the material used in construction of many termite nests (Lee and Wood 1971b); carton is highly absorbent and should not constitute any barrier to water uptake through burrow walls.

6. Protection of Nests and Burrows from Flooding

It is important for the survival of subterranean termites and ants that their nests do not fill with water during periods of surface ponding. For mound nests that provide living space and food storage that are mainly above ground this is probably not an important problem. There are several techniques used by ants and termites that live below the ground surface to avoid flooding of their nests. Many termite species seal surface openings except when they are in use for foraging or the exit from the nest of winged adults (alates). Ants commonly construct a funnel-shaped barrier of soil around the nest entrance which, although it forms a small catchment from which the water must drain into the nest, acts as a dam to divert water flowing across the surface away from the nest entrance. A mound of the ant *Formica cinerea*, 40 cm high and 140 cm in diameter, was shown by Denning et al. (1977) to have 16% by volume of macropores, giving it a saturated conductivity of 100 to 400 m per day, which was at least a thousand times the conductivity needed to handle the ponded water that would result from the maximum expected 24-hour precipitation. An extreme case of modification of the soil below termite mounds was described by Boyer (1973, 1975a, 1975b) for very large mounds of *Macrotermes bellicosus* in the Central African Republic; pedogenesis and regional hydrology were shown to be modified over a very large savanna region, with a zone of soil below each of the mounds modified so that it functioned as a conduit for water to the permanent water table.

III. Spatial Pattern of Biopores

The areal pattern of burrows in the horizontal plane can influence the variability of ponded infiltration (Smettem and Collis-George, 1985a) and the probability of root interception of a burrow (Dexter, 1986).

The areal pattern of burrows may generally be described by one of three statistical models. Burrows with a regular spacing may be described by the binomial model obtained from the expansion of $(q + p)^k$ where p is the probability of an event occurring one way and q is the probability of the same event occurring in an alternative way. If q and p are constant in k independent trials then the expansion is valid and the individual terms follow from:

$$P(x) = \{k!/x!(k-x)!\}\ q^{k-x}\, p^x \qquad (1)$$

where P(x) is the probability of x individuals in a sampling unit and k! is k factorial. In this case, k is the maximum possible number of burrows a sampling unit could contain and may be approximated from $k = \bar{k} = \bar{x}^2 / \bar{x}\text{-}s^2$, where $\bar{x}$ is the mean number of burrows in a sampling unit and s^2 is the variance. p and q are estimated from $p = \bar{x} / \bar{k}$; $q = 1\text{-}p$. Ecologically, such burrow systems would tend to be formed by territorial animals with a defined foraging area away from the burrow. Burrow systems may also be randomly distributed or clumped (contagious). Mature earthworms are generally randomly distributed (Edwards and Lofty, 1977) and it is plausible to expect their burrow systems to be described similarly. Smettem and Collis-George (1985a) found cylindrical pores >0.5 mm diameter were predominantly randomly distributed at 10 cm depth under a pasture near Sydney, Australia. They used nearest neighbor methods to determine the areal pattern of the burrows. The resulting random distribution was described using the Poisson series given by:

$$P(x) = e^{-\lambda}\, \lambda x/x! \qquad x = 0,1,2....n \qquad (2)$$

where λ is the Poisson parameter, estimated from the sample mean or the variance ($\lambda = \bar{x} = s^2$).

In South Australia greater numbers of *A. caliginosa* were observed around the roots of wheat plants than in the intervening row spaces (Rovira et al., 1987). This could result in a clumped areal pattern of burrows with reduced burrowing activity in the inter-row spaces. The resulting contagion of the distribution could be described by the negative binomial distribution given by the expansion of $(q - p)^{-k}$ where $p = \bar{x}/k$ and $q = 1 + p$. The parameters of this distribution are the arithmetic mean $\bar{x}$ and the exponent k that is related to the areal pattern of the burrows. Cressie (1991) reviews these and other point-process models for clumped distributions.

Describing biopores as points within a given cross-sectional area of the soil "matrix" is a highly simplified but useful morphological description that has implications for both water entry and root penetration of subsoils. These topics are developed in a later section.

IV. Effects of Biopores on Water Movement

Soils permeated by earthworm burrows or other macropores may be envisaged as a two-domain flow system, with flow in the smaller soil 'matrix' pores dominated by capillary effects, and that in the larger pores as rapid movement of essentially "free" water (Germann and Beven, 1981; Bouma, 1981; White, 1985; Smettem and Collis-George, 1985b). Water entry into the burrow system usually occurs when a positive pressure head (ponding) develops at the soil surface, or over a layer of low permeability within the profile. Seepage into the burrow may also occur in unsaturated soil if the pressure at some point on the walls reaches that of the local soil atmosphere (Philip, 1987).

A plethora of studies have demonstrated the influence of earthworms on infiltration under ponded conditions. Slater and Hopp (1947), Hopp and Slater (1948), Teotia et al. (1950), Guild (1952, 1955), Scharpenseel and Gewehr (1960) and Carter et al., (1982) found that water entered the soil between two and ten times faster in soils with earthworms than in soils lacking them. Bouché (1971) poured 100 litres of water down one earthworm burrow without it overflowing; he showed that the burrows intersect and that water could penetrate through a network of burrows. Ehlers (1975) demonstrated that an increase in the number of earthworm burrows under zero till compared to conventional cultivation resulted in a six fold increase in infiltration over a five minute period. Tisdall (1978) demonstrated that increasing earthworm numbers from 150 m^2 to 2000 m^2 in irrigated orchards in Victoria, Australia resulted in increased infiltration rates from 50 mm in 83 minutes at the untreated sites to 50 mm in 1 minute at the treated sites.

Ehlers (1975) and Smettem and Collis-George (1985b) measured infiltration directly into burrows and correlated infiltration with burrow radius. Smettem and Collis-George (1985c) then predicted distributions of ponded infiltration rates by combining this information with the Poisson model for the areal pattern of pores described by Eq. (2). Their study showed that a "Representative Elemental Area" (REA) for ponded infiltration measurement followed from the Central Limit Theorem of classical statistics when the areal scale of a sample ring included at least ten of the most conductive class of biopores. For rings smaller than this REA, infiltration distributions became increasingly skewed as ring size decreased. This is because the probability of obtaining no macropores within a sample ring increases with decreasing ring size, leading in turn, to an increasing number of measurements that reflect only "matrix" infiltration properties. The implications of this study are that comparative measurements must be made at a suitable REA when assessing the effects of soil biopores on soil properties. An area of at least 1 m^2 appears to be appropriate (Beven and Clarke, 1986).

Under natural rainfall, or sprinkler irrigation, water that begins to pond on the soil surface will flow towards microtopographic hollows. The filling of a particular burrow will therefore depend upon the interaction between surface microtopography and the burrow vent at the soil surface. Trojan and Linden (1992) found that the deepest penetration of a dye tracer coincided with burrows located at the lowest elevations in surface microdepressions. Until the entire soil surface is ponded, the amount of water that enters a particular burrow will be influenced strongly by the surface "catchment" of the burrow vent. An important consequence of this is that the "catchment" of the burrow will tend to dictate how much water percolates through the burrow and in turn, how much surface chemical is transported through the burrow to depth (Edwards et al., 1992).

Douglas et al. (1980) studied the continuity of "transmission" pores, i.e., pores >50 μm diameter, from A to B horizons of a clay soil (about 50% clay), comparing plowed with direct-drilled treatments. Mean hydraulic conductivity (K) at the interface of the A and B horizons in direct drilled treatments was found to be twice that in plowed treatments. Soil cores from direct-drilled soils that included the A-B horizon transition were found to be perforated by continuous, vertical cylindrical earthworm burrows, 1-8 mm in diameter. When these channels were artificially blocked at each end K decreased by 80% and, although the pores were irregularly spaced and constituted only a small proportion of the total pore space, it was apparent that they provided an enhanced link for transmission of water (and gases) into the B horizons of direct-drilled soils. Total porosity and volume of transmission pores was shown to be higher in plowed than in direct-drilled soils, but earthworm burrows in the direct-drilled soil were considered to be particularly significant because of their continuity from the surface to the B horizon.

Smettem (1987) pointed out that measurements of flow through soil cores may give unrealistically high estimates for field hydraulic conductivities because macropores may be continuous throughout the core, whereas in the field they are likely to terminate somewhere within the soil profile. To account for this difference a distinction was made between "dynamic" and "passive" macropore flow. Passive flow referred to the condition where filled macropores simply respond to conditions in the surrounding soil, whereas dynamic flow would tend to be controlled by the radius of the macropore according to Poiseuille-type behavior (Germann and Beven, 1981)

$$Q = Q^* \pi \, r^4 \tag{3}$$

where Q is the volume flux ($l^3 \, t^{-1}$), Q* is a constant and r is the pore radius (L).

Dynamic flow is observed in macropores as they fill and in burrows exiting from stream banks or emptying into larger structural features. Bouma et al. (1982) observed both types of flow through 6 mm diameter burrows in a silt loam subsoil in the Netherlands. For burrows that terminated at 1.6 m, infiltration rates averaged 140 $cm^3 \, min^{-1}$, however for burrows ending within 0.5 m of the soil surface in highly permeable mole burrows, rates averaged 400 $cm^3 \, min^{-1}$. The volume of water stored within filled macropores has also recently been referred to as 'internal catchment' (Van Stiphout et al., 1987).

Smettem (1992) discussed the relation of earthworms to soil hydraulic properties and showed how some relatively simple equations could be used to assist the development of management techniques aimed at manipulating earthworm populations to achieve desired hydrologic behavior. The first analysis applies to the situation where water begins to enter the earthworm burrows and the flow process is no longer one-dimensional because infiltration now occurs at the soil surface and through the walls and base of the burrows. During this period, the surface area of macropores exposed to the soil surface, A_m (L^2), may be estimated from sorptivity measurements at two different supply potentials, one ponded and the other unsaturated, using (White et al., 1992)

$$A_m/(\pi r^2) = (S_1/S_2)/(1+ \Delta\psi /\lambda_c)^{1/2} - 1 \quad (4)$$

where S_1 is the ponded sorptivity, S_2 is the sorptivity measured at a slight tension that excludes the macropores from the flow process (e.g., using a disc or tension infiltrometer), πr^2 is the infiltrometer source area, $\Delta\psi$ is the supply potential difference between the two readings and λ_c is the capillary length scale, which is related to the sorptivity S_2 by

$$\lambda_c = bS_2^2/ \Delta\theta \, (K_2-K_n) \quad (5)$$

where the subscripts 2 and n refer to the supply and initial conditions respectively, $\Delta\theta$ is the volumetric water content difference, K is the hydraulic conductivity, with K_2 measured at a tension corresponding to S_2 and b is a parameter that depends on the shape of the soil-water diffusivity function and is constrained to $0.5 < b < \pi/4$. A typical "average" value is 0.55. Normally K_n is neglected because $K_2 >> K_n$. All parameters in Eq. (5) may be obtained using the disc permeameter (see White and Sully, 1987; Smettem and Ross, 1992 and Smettem, 1992 for examples).

White et al. (1992) point out that Eq. (4) is an approximation because it assumes that the macropores fill much faster than the rate of sorption of water into the soil matrix and that S_1 and S_2 are measured within the range of tension saturation.

Once the earthworm burrows are filled, quasi-steady-state flow conditions are rapidly achieved and the analogy to flow out of a borehole permeameter can be used to estimate the flow out of the burrow into the surrounding unsaturated soil. The solution of Philip (1985) is appropriate

$$Q = H\{4.117H/[ln(2H)-1]\} + \{[4.028 + 2.517H^{-1}]/[A\,ln(2H)]\} \quad (6)$$

where

$$Q = q/(K_2a^2) \quad (7)$$
$$H = h/a \quad (8)$$
$$A = 1/2\lambda_c^{-1} \quad (9)$$

and q is the volume flux (L^3T^{-1}), h is the burrow length (L) and a is the burrow radius (L).

The terms on the right hand side of Eq. (6) describe the flow due to pressure and capillary absorption respectively. The influence of the initial soil water content is reflected in λ_c which is larger when the soil is dry. Again, all the hydraulic parameters in Eq. (6) may be obtained using the disc permeameter. Additional morphological information on the length and radii of conducting burrows is also required. To date, this has generally been obtained by excavation of individual burrows filled with gypsum slurry (Bouma et al., 1982) or paraffin wax (Smettem, 1986).

Once the dimensionless flux has been determined using Eq. (6), the volume flux follows by rearrangement of Eq. (7), with K_2 obtained from the disc permeameter

$$q = QK_2a^2 \tag{10}$$

Eq. (10) can be coupled to a spatial parameter (e.g., the Poisson parameter) in order to areally estimate the contribution of water-filled biopores to infiltration prior to the time when the expanding wetting bulbs from adjacent biopores come into contact. A similar procedure was adopted by Beven and Clarke (1986).

V. Identification of Functional Biopores

Using fluorescent dyes, Omoti and Wild (1979) found that most of the burrows they observed at 0.5 m conducted water from the surface. Smettem and Collis-George (1985a) however found a discrepancy at 0.1 m between pores stained with the fluorescent anionic dye lissamine yellow ff and the cationic dye methylene blue. On a number of soil peels lissamine was found in 86-100% of the pores >0.5 mm diameter, whereas methylene blue was only observed in 41-46% of these pores. This discrepancy was most marked for pores less than 2.5 mm diameter and suggests that the flow rate in these pores was considerably slower than the rate of methylene blue absorption. Lissamine yellow may take about 24 hours to reach adsorption equilibrium (Smettem and Trudgill, 1983) and also has a much lower adsorption retardation factor than methylene blue.

Although the presence of dye on the walls of a biopore can be taken as indicating a flow pathway, the converse is not necessarily true. The observations of Smettem and Collis-George (1985a) are one example where interpretation of methylene blue staining would underestimate the number of conducting pores at 0.1 m depth. Similar underestimation can occur if a "perched" water table develops ahead of the dye front over a confining layer. "Free" water will then enter subsoil macropores which may remain unstained and hence undervalued in terms of their functionality. A further limitation is that it is not possible to determine whether fluid has been conducted by the whole cross-sectional area of

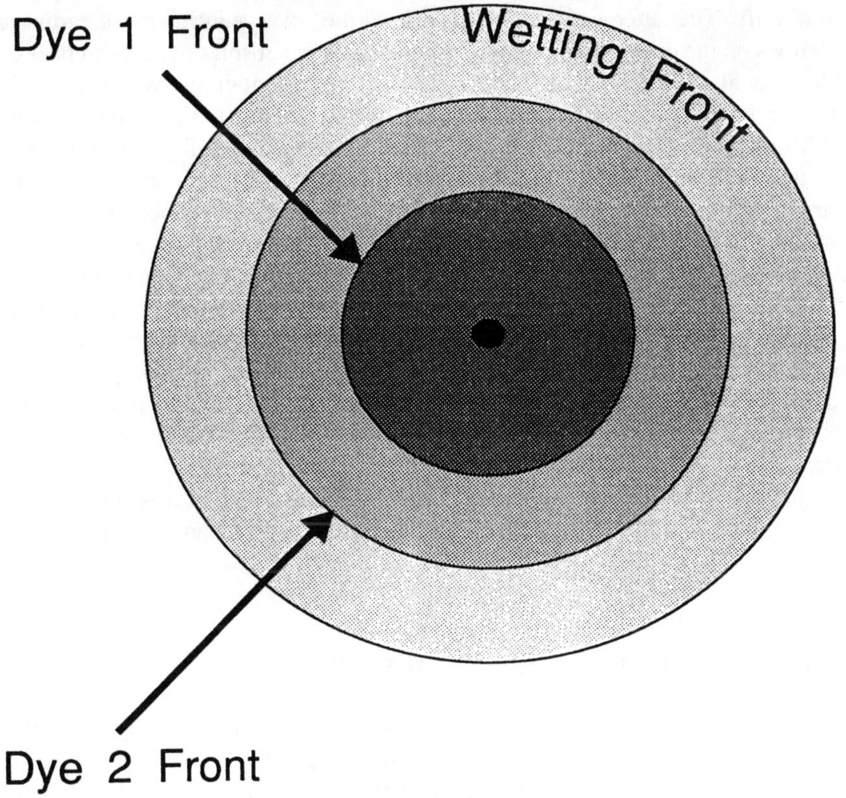

Figure 1. Schematic cross-section of water and dye movement away from a macropore into unsaturated soil. Both dyes are adsorbed, with dye 1 adsorbed more strongly than dye 2. The cross-sectional area of staining is strongly dependent on dye adsorption characteristics.

the biopore or restricted to annular flow. In both cases the walls of the biopore will be stained.

The practice of interpreting the cross-sectional area that has been stained as the conducting area is also questionable. Consider the case of water flowing through a single cylindrical biopore in an otherwise unsaturated "matrix". In cross-section, water absorption into the surrounding soil will occur as an expanding circle of wetting. If the water contains a dye tracer then (unless there is no adsorption, which is unlikely) the front of the dye will lag some distance behind the water front. Furthermore, if two dyes with different adsorption distribution coefficients are considered, then the stained cross-sectional area will be different for the two dyes. This is illustrated schematically in Figure 1.

Gypsum slurries or plaster of Paris have been used by Bouma et al., (1982), Fitzpatrick et al. (1985) and Urbanek and Dolezal (1992) to study structural continuity. The latter authors also quantified the positive influence of earthworm burrows on the preservation of high permeability in subsoil drain trench backfills. Moran et al. (1989) have used a low viscosity resin containing two different dyes in an attempt to impregnate the soil pore space in field condition. In the method of Moran et al. (1989), the first dye is used when resin is infiltrated into the soil in the field by means of a constant-head drip infiltrometer. The resin is formulated to cure in about 24 h and partially impregnated specimens are then returned to the laboratory where impregnation is completed under vacuum with resin containing a different dye. This three-component approach yields more information from a vertical section than from a similar two-component impregnation (Moran and McBratney, 1992) and provides a useful partitioning of conducting and non-conducting pore space. Clearly, despite progress, the identification of conducting biopores remains an area that requires research, particularly in relation to subsoil management. Given that the length and surface area of conducting biopores are critical morphological parameters for simple predictions of water flow (Smettem, 1992), effort invested in this area should be of practical benefit for developing biopore-related subsoil management strategies.

VI. Effects of Biopores on Root Growth

The ability of plant roots to penetrate deep into the soil is an important factor in drought tolerance. Elkins et al. (1977) investigated the penetration of roots of crop plants in sandy soils in Alabama. If the roots were confined to the top 15 cm of the soil, the crop suffered water stress after three days without rain, whereas, when roots could penetrate to 100 cm, the crop suffered water stress only after 20 days without rain. Roots that grow down through a seed bed and then encounter an impenetrable compact subsoil layer grow horizontally along the top of the subsoil layer until the seed bed dries out and the roots wilt or until they find a path of reduced resistance, where they are able to penetrate the compact layer (Dexter 1986).

Burrows of soil animals especially those of earthworms are more continuous, more stable, and less tortuous than macropores created by tillage operations; they provide preferred pathways for water and air movement and for the penetration of roots (Shipitalo and Protz, 1987; Francis et al., 1988). This is particularly important where no-tillage or direct drilling practices lead to increased mean values of bulk density and soil strength, thereby limiting root elongation (Hamblin and Tenant, 1979; Chan et al., 1987). The buildup of biopores under zero tillage can be used by the roots of subsequent crops as paths of low mechanical impedance (Ehlers et al., 1983). Edwards and Lofty (1978) also observed a close correlation between root distributions and zones of activity of earthworms and other soil fauna.

In contrast, Logsdon and Allmaras (1991) and Logsdon et al. (1992) generally observed that only a small portion of the roots of corn, soybean, alfalfa and oat crops were associated with worm burrows in an Alfisol and Mollisols ranging in texture from fine sandy loams to silt loams and clay loams.

Dexter (1986) has modeled the probability of roots entering biopores after lateral deflection due to encounter with a strong untilled subsoil. He compared the model with experimental data obtained from growing roots of pea and wheat through beds of aggregates to encounter an artificial impenetrable subsoil at random positions. The roots were deflected horizontally until they encountered vertical holes drilled in the subsoil. Regular or randomly spaced arrays of holes with varying diameters were used in the experiments and the distances that randomly deflected roots would have to travel before encountering holes were studied as functions of hole diameter and density. For a range of hole densities, H, and diameters, D_h, the probability, Pe, of a root encountering at least one hole in a random array of holes was described by the equation:

$$Pe = 1 - e^{-1.23H(x/Dh)} \qquad (11)$$

where x/D_h is a dimensionless distance and H is a dimensionless hole density. H is related to hole diameter D (mm) and the number of holes per m^2 , p, by:

$$H = (\pi p D_h^2/2) \times 10^{-6} \qquad (12)$$

An example of the simulation is shown in Figure 2. After comparing the experimental results with the computer simulations, Dexter concluded that the numbers of roots encountering holes within certain distances were not significantly different in practice from those simulated on the basis of random chance.

The importance of this work is that it offers a method for estimating the probability of roots intercepting soil biopores and entering the subsoil using easily obtained morphological information on the spatial pattern, diameter and spatial intensity (number per area) of biopores.

In soils that have compact subsoil horizons roots are commonly clumped together in biopores. Passioura (1991) considered this situation, where the biopores are earthworm burrows. The burrows are usually so widely spaced that nominally available water may be poorly accessible, and roots may fail to extract water, especially from deep in the subsoil, even though substantial lengths of roots are present at this depth. A clump of roots in a biopore may be regarded as analogous to a single root that has access to water from a cylinder of soil surrounding the biopore. In this situation it becomes important to understand the extent to which the roots are clumped into biopores, rather than simply measure the root length density. Methods need to be devised to measure the spatial intensity of biopores required to allow the clumped root system to effectively extract subsoil moisture. Furthermore, we need to know more about the hydraulic continuity between the roots of different species and the soil surrounding

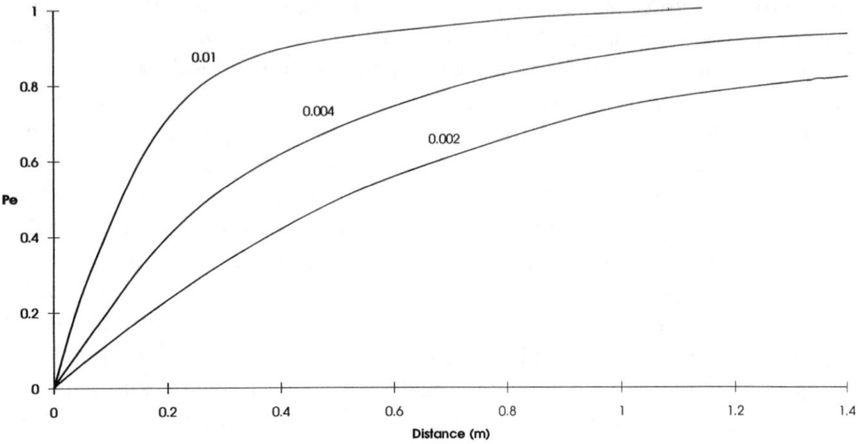

Figure 2. Examples of results from computer simulations of roots encountering random arrays of cylindrical holes. Probabilities, Pe, of a root encountering at least one hole of diameter 2 mm within a distance x (mm) are shown. Numbers on curves are values of the proportion, H, of the area of the subsoil surface which is occupied by holes. (Redrawn from Dexter, 1986; reproduced with permission of Plant and Soil.)

biopores if we are to determine whether extraction patterns are likely to differ between plant species.

VII. Influence of Burrowing Soil Animals on Dispersal of Microorganisms

Soil microorganisms, fungal propagules, protozoa, nematodes, micro-arthropods and other small soil organisms are ingested by earthworms and there is evidence that many are digested and are an important part of earthworms' diet. Others thrive and multiply as they pass through the gut, while some, especially those that have spores, cysts or other resistant life stages, pass unchanged through the gut, to be excreted in casts. A review of the subject may be found in Lee (1985).

Usually, the maximum concentration of organic matter is in the topsoil close to the surface and this is where microorganisms are most numerous. For example, Schnurer et al. (1986) showed that 73% of total lengths of fungal hyphae and 79% of the total bacterial biomass were in the topsoil layers of Swedish soils under four systems of cropping. In some circumstances, especially in soils that are subject to drought, concentrations of important plant symbionts are found, associated with roots that penetrate deep into the subsoil to sources of water. Virginia et al. (1986) studied the root systems and symbiotic N-fixing rhizobia of the mesquite (*Prosopis glandulosa*) in the Sonoran Desert, and showed that the population density of the rhizobia was higher in deep soil (2.5 to 5 m depth) than in overlying soil horizons, with a maximum at about 5 m. Roots infected with mycorrhizal fungi were also recovered from depths down to 4.8 m, and it was concluded that the root-mycorrhizal association would be extracting nutrients from the surrounding soil. The mechanism that moved these microorganisms to such depths was not investigated, but it seems likely that they would be associated with biopores and other macropores.

The burrows of earthworms, termites, ants, and other soil macrofauna must, when air-filled, also provide pathways for smaller soil animals that cannot themselves make burrows to move into and within soil profiles, and when water-filled to carry small soil organisms passively down soil profiles; the ecological and perhaps plant pathological significance of these pathways still requires research.

Evans and Miller (1988, 1990) noted phosphorus deficiency in maize growing in plowed soil and its absence in maize grown in uncultivated soil and attributed this difference to different levels of mycorrhizal infection of the roots. Passioura (1991) proposed that the difference in infection may relate to the clumping of roots in biopores, which are particularly well developed in uncultivated soil. Mycorrhizal fungi would be concentrated in the biopores, so enhancing the efficiency of establishment of symbiotic relationships with plant roots, especially in the uncultivated soil, and facilitating the uptake of phosphorus.

Reddell and Spain (1991a, 1991b) have demonstrated that spores of mycorrhizal fungi and infective propagules of the N-fixing symbiotic actinomycete *Frankia* remain viable after passage through the gut of earthworms and deposition in casts. Mycorrhizal fungi were found to be concentrated in the casts of earthworms of a range of ecological groups, compared with numbers in surrounding soil. It was concluded that earthworms are effective vectors of these microorganisms, vertically within soil profiles, and horizontally over short distances. Particular attention was drawn to the pan tropical endogeic earthworm *Pontoscolex corethrurus*, which deposits casts and presumably spores of mycorrhizal fungi in its burrows. These burrows are known to be preferred root growth pathways and roots growing in them must be particularly susceptible to mycorrhizal infection.

VIII. Management of Soil Organisms to Increase Macroporosity

Soil organisms play an important part in maintaining and enhancing soil fertility, and hence in making agricultural practices sustainable (Greenland, 1991). Land development for agriculture involves clearing of the original vegetation, usually some cultivation, followed by monoculture or at least a change to ecosystems with less plant species, different from those in the original systems, and a reduction in habitat complexity compared with that in the original undisturbed state. Changes include destruction or drastic modification of some niches for soil organisms, and the disappearance of some species, especially those whose niche widths in the original ecosystems were narrowly limited (Lee, 1991). Earthworms, which are particularly significant in establishing and maintaining soil structure and macroporosity that favor plant growth, are usually eliminated or at least much reduced in diversity and abundance. They may or may not be replaced by exotic species. The most successful exotic species of earthworms comprise a small number of species of originally European Lumbricidae and South American Megascolecidae in temperate regions of the world, and of Glossoscolecidae and Megascolecidae, originally from tropical South and Central America, Africa and South-East Asia, in tropical regions. They are opportunistic, typical *r*-selected species, with a capacity to expand or modify their apparent niche width to adapt to new circumstances (Lee, 1987). The chances of replacement and the species that move into the empty niches depend only on the availability in adjacent soils of opportunistic colonising species that can move unaided or may be introduced accidentally by man. These are not necessarily the species likely to provide the maximum benefits for plant growth in new agricultural ecosystems, so there is an opportunity to select species and to manage their introduction.

Evidence to date suggests that an ideal earthworm population would include species that make deep burrows and so provide a network of macropores that penetrate into the subsoil (Smettem, 1992). The ideal population should include one or more anecic species that feed and deposit their casts at the soil surface, so removing and burying plant litter, and one or more endogeic species that feed on organic matter below the soil surface and that make more horizontal burrows, so facilitating lateral movement of water and gases within the soil (see Lee, 1991). It may not be necessary to select separate species for each of these roles, as some of the most widespread of species, such as *Aporrectodea caliginosa* and *Pontoscolex corethrurus,* are able to expand their 'normal' niches to provide beneficial effects that might be expected of several species in their original environment (Lee, 1987; 1991). There are some 3,000 species of earthworms from which selection can be made, although there are not more than about 12-15 species whose ecology and behavior are known in sufficient detail to make selections that could be predicted to have a good chance of success.

In many countries, successful earthworm introduction programs have been followed by increases in infiltration rates and plant productivity. For example, introduction of *A. caliginosa* to New Zealand pastures (Stockdill, 1966; 1982) resulted in about 100% enhancement of infiltration rate, with 28% increase in pasture production 10 years after the introduction, while Hoogerkamp et al. (1983) recorded a 10% increase in pasture production on polder soils 10 years after the introduction of *A. caliginosa.* The earthworms introduced have usually been species that are active near the soil surface, and techniques for their introduction over substantial areas of farmland, using turfs taken from farmlands where the desired species are common, are now established. Techniques for obtaining and introducing large numbers of deeper burrowing species are not well developed. Springett (1983, 1985) introduced *Aporrectodea longa,* a deep burrowing lumbricid, to pasture plots with resident populations of *A. caliginosa* and *Lumbricus rubellus*, which are active mainly within the top 10 cm of the soil profile. The worms were collected by digging and hand-sorting from pastures with high populations of *A. longa,* were matched for size and proportion of mature specimens, and were placed on the surface of the plots at a rate of 150 m^{-2} . One year after introduction of *A. longa* it was observed that:

- agricultural lime added at the soil surface or in a layer at 10 cm depth was transported to depths below 10 cm more rapidly than in plots without *A. longa*;
- the mean biomass of roots at 15-20 cm depth was 92% more ($p < 0.01$) than in plots without *A. longa*;
- total porosity at 15-20 cm increased from 56% in plots without *A. longa* to 64% ($p < 0.05$) in plots with *A. longa*;
- water infiltration rate at 20-25 cm increased from 96 mm h^{-1} in plots without *A. longa* to 144 mm h^{-1} in plots with *A. longa*.

Springett's work illustrates the potential value of introducing deep-burrowing species to improve subsoil macroporosity by supplementing the effects of species that live closer to the surface. Successful introduction of the deep burrowing earthworm *Eisenia djungarica,* collected from apple tree forests of the Djungar Alatau mountains in Kazakhstan to apple orchards in Alma Ata, was reported by Ghilarov and Perel (1984). The worms were collected by hand and introduced into 0.2 × 0.2 × 0.2 m pits, five individuals per pit. Four years after introduction, populations up to 10 m^{-2} were found over areas of about 150-200 m^2 around the sites of the introduction pits. No information is available on their effects on soil properties.

Curry (1988) suggested modifications of the method of transplanting sods of turf containing shallow-dwelling species that was developed by Stockdill in New Zealand, including the taking of larger sods, the taking of sods at times of the year when juvenile stages of the deep-burrowing species can be expected to be numerous and to be active near the soil surface, and the possibility of boosting populations in the donor soil by liberal applications of farmyard manure or similar materials prior to transplanting.

Possibilities for the management of termite and ant populations to enhance soil fertility, and the role of their burrows in promoting soil aeration and the infiltration of water into deep soil horizons, have not been investigated. Most subterranean termites are small and insignificant animals that come to the surface to forage for food mainly at night. Large populations can go unnoticed. For example, it is not generally recognized that over much of the wheat producing areas of South Australia *Amitermes* spp. live in diffuse burrow systems that penetrate to depths of 0.5 m or more in the soil. In late spring or early summer, when the adult termites emerge at night to make their mating flights, the soil surface over large areas is strewn with their discarded wings, and it is apparent that very large numbers of termites have been active; however, no research has been done on the density or distribution of termite populations or on the distribution of their burrows in the soil and their possible pedological significance. Sampling of subterranean termites for population studies is so labor intensive that such studies are probably not practicable, but assessment of the pedological significance of the termites' burrows would not be so difficult, and it might indicate management practices to enhance any beneficial effects.

IX. Conclusions and Recommendations

The management of subsoil problems by manipulating the activity of soil organisms to produce beneficial effects, particularly soil biopores, has potential but still requires a substantial investment of integrated research. We have still not secured good descriptions of how water and chemicals are distributed in subsoils, nor how plant roots are distributed and how they extract water and nutrients. In particular, the significance of root clumping in soil biopores needs to be assessed in systems where extraction of subsoil water is an important constraint to crop production. Allied to this, the environmental and soil factors that cause roots to concentrate in subsoil burrows under some circumstances but not in others need to be better understood.

We have good biological understanding of how and why soil animals form burrows and we understand some of the physical constraints to burrow formation. Soil morphological descriptions can set reasonable upper (total bioporosity) and lower (stained bioporosity) limits for the functional biopore space but work is still needed to improve descriptions of three-dimensional continuity and better link morphological descriptions to practical physical attributes. There is also a need for good techniques to observe growing roots in thin section so that the interaction between roots and biopores can be more clearly investigated.

Methods for managing soil organisms are still at an early stage of development. Broad-acre applications pose considerable technical problems and are not a viable option until research can provide methods of mass culture for deep burrowing species or methods of collection that are not so labor-intensive or environmentally destructive as digging. Recently, Butt et al. (1992) have developed a successful system for mass rearing of *Lumbricus terrestris,* so some progress is being made. In small-scale high-return ventures, however, it does appear to be within our capability to manipulate populations and achieve cost-effective benefits, particularly in horticultural activities and also at a larger scale under improved pasture management systems.

Given the wealth of evidence that some soil organisms confer agronomically beneficial effects on soils, either directly or indirectly, there is an urgent need to develop cost-effective management strategies that enhance the beneficial effects of useful soil organisms, as well as to continue the development of integrated chemical and biological control methods to reduce the damage done by pest organisms.

References

Barley, K.P. 1959. Earthworms and soil fertility. IV. The influence of earthworms on the physical properties of a red-brown earth. *Aust. J. Agric. Sci.* 10:371-376.

Beven, K. and P.F. Germann. 1982. Macropores and water flow in soils. *Water Resour. Res.* 18:1311-1325.

Beven, K. and R.T. Clarke. 1986. On the variation of infiltration into a homogenous soil matrix containing a population of macropores. *Water Resour. Res.* 22:383-388.

Bhatnagar, T. 1975. Lombriciens et humification: un aspect nouveau de l'incorporation microbienne d'azote induite par les vers de terre. p.169-182. In: G.Kilbertus, O. Reisinger, A. Mourey and J.A. Consela da Fonseca (eds.) *Biodégradation et Humification.* Pierron,Sarreguemines.

Bolton, P.J. and J. Phillipson. 1976. Burrowing, feeding, egestion and energy budgets of *Allobophora rosea* (Savigny) (Lumbricidae). *Oecologia* (Berlin) 23:225-245.

Bouché, M.B. 1971. Relations entre les structures spatiales et fonctionelles des ecosystèmes, illustrees par le rôle pédobiologique des vers de terre. p. 189-209. In: C.Delamere Deboutteville (ed.) *La Vie dans les Sols*, Gauthiers Villars, Paris.

Bouché, M.B. 1975. Action de la faune sur les états de la matière organique dans lesècosystemes. p.157-168 In: G. Kilbertus, O. Reisinger, A. Mourey and J.A. Consela da Fonseca (eds.) *Biodégradation et Humification.* Pierron, Sarreguemines.

Bouché, M.B. 1977. Strategies lombriciennes. In: U. Lohm and T. Persson (eds.) *Soil Organisms as Components of Ecosystems.* Ecological Bulletins (Stockholm) 25:122-132.

Bouché, M.B. and R.H. Gardiner. 1984. Earthworm functions. VIII. Population estimation techniques. *Rev. Ecol. Biol. Sol.* 21:37-63.

Bouma, J. 1981. Soil morphology and preferential flow along macropores. *Agric. Water Mgt.* 3:235-250.

Bouma, J. 1994. Effect of soil structure, tillage, and aggregation upon soil hydraulic properties. *Adv. Soil Res.* (in press).

Bouma, J., C.F.M. Belmans and L.W. Dekker. 1982. Water infiltration and redistribution in a silt loam subsoil with vertical worm channels. *Soil Sci. Soc. Am. J.* 46:917-921.

Boyer, P. 1973. Actions de certains termites constructeurs sur l'evolution des sols tropicaux. I. Les termites et le sol. *Annal. Sci. Nat. Zool. Paris* Ser.12. 15:329-428.

Boyer, P. 1975a. Action de certains termites constructeurs sur l'evolution des sols tropicaux. II. Etude particulière de trois termitières de *Bellicositermes* et de leur action sur les sols tropicaux. *Annal. Sci. Nat. Zool. Paris* Ser.12. 17:273-446.

Boyer, P. 1975b. Action de certains termites constructeurs sur l'evolution des sols tropicaux. III. Les differents aspects de l'action de *Bellicositermes* sur les sols tropicaux. *Annal. Sci. Nat. Zool. Paris* Ser.12. 17:447-504.

Butt, K.R., J. Frederickson and R.M. Morris. 1992. The intensive production of *Lumbricus terrestris* L. for soil amelioration. *Soil Biol. Biochem.* 24:1321-1325.

Carter, A., J. Heinonen and J. deVries. 1982. Earthworms and water movement. *Pedobiologia.* 23:395-397.

Chan, K.Y., J.A. Mead and W.P. Roberts. 1987. Poor early growth of wheat under direct drilling. *Aust. J. Agric Res.*, 38: 791-800.

Cressie, N. 1991. *Statistics for Spatial Data.* Wiley Interscience, New York. 900 pp.

Curry, J.P. 1988. The ecology of earthworms in reclaimed soils and their influence on soil fertility. p 251-261. In: C.A. Edwards and E.F. Neuhauser (eds). *Earthworms in Waste and Environmental Management.* SBP Academic Publishing, The Hague, Netherlands.

Curry, J.P. and J.A. Good. 1992. Soil faunal degradation and restoration. *Adv. Soil Res.* 17:171-203

Darlington, J.P.E.C. 1982. The underground passages and storage pits used in foraging by a nest of the termite Macrotermes michaelseni in Kajiado, Kenya. *J. Zool.*(London) 198:237-247.

Denning, J.L., F.D. Hole and J. Bouma. 1977. Effects of *Formica cinerea* on a wetland soil on West Blue Mound, Wisconsin. p.276-87 In: C.C.B. DeWitt and E. Soloway (eds.) *Wetlands, Ecology, Values and Impacts.* University of Wisconsin-Madison, Madison. USA.

Dexter, A.R. 1978. Tunnelling in soil by earthworms. *Soil Biol. Biochem.* 10:447-449.

Dexter, A.R. 1986. Model experiments on the behaviour of roots at the interface between a tilled seed-bed and a compacted sub-soil. III. Entry of pea and wheat roots into cylindrical biopores. *Plant Soil* 95:149-161.

Douglas, J.T., M.J. Goss and D. Hill 1980. Measurements of pore characteristics in a clay soil under ploughing and direct drilling, including use of a radioactive tracer (144Ce) technique. *Soil Tillage Res.* 1:11-18.

Edwards, C.A. and J.R. Lofty. 1977. *Biology of earthworms.* 2nd ed. Chapman and Hall, London.

Edwards, C.A. and J.R. Lofty. 1978. The influence of arthropods and earthworms upon root growth of direct drilled cereals. *J. Appl. Ecol.* 15:789-795.

Edwards, W.M., M.J. Shipitalo, W.A. Dick and L.B. Owens. 1992. Rainfall intensity affects transport of water and chemicals through macropores in no-till soil. *Soil Sci. Soc. Am. J.* 56: 52-58.

Ehlers, W. 1975. Observations on earthworm channels and infiltration on tilled and untilled loess soil. *Soil Sci.* 119:242-249.

Ehlers, W., U. Kopke, F. Hesse and W. Bohm. 1983. Penetration resistance and root growth of oats in tilled and untilled loess soil. *Soil and Tillage Res.* 3:261-275.

Elkins, C.B., R.L. Haaland and C.S. Hoveland. 1977. Grass roots as a tool for penetrating soil hardpans and increasing crop yields. *Proc. 34th Southern Pasture and Forage Crop Improvement Conf.* Auburn, Alabama 21-26.

Evans, D.G. and M.H. Miller. 1988. Vesicular-arbuscular mycorrhizas and the soil-disturbance-induced reduction of nutrient absorption in maize. *New Phytologist* 110:67-74.

Evans, D.G. and M.H. Miller. 1990. The role of the external mycelial network in the effect of soil disturbance upon vesicular-arbuscular mycorrhizal colonization of maize. *New Phytologist* 114: 65-71.

Fitzpatrick, E.A., L.A. MacKie and C.E. Mullins. 1985. The use of plaster of Paris in the study of soil structure. *Soil Use Manag.* 1:70-72.

Francis, G.S., K.C. Cameron and R.A. Kemp. 1988. A comparison of soil porosity and solute leaching after six years of direct drilling or conventional cultivation. *Aust. J. Soil Res.* 26:637-649.

Germann, P.F. and K. Beven. 1981. Water flow in soil macropores 3. A statistical approach. *J. Soil Sci.* 32:31-39.

Ghiralov, M.S. and T.S. Perel. 1984. Transfer of earthworms (Lumbricidae,Oligochaeta) for soil amelioration in the USSR. *Pedobiologia.* 27:107-113.

Greenland, D.J. 1991. Foreword. p. xvii-xviii. In: D.L. Hawksworth (ed.) *The Biodiversity of Microorganisms and Invertebrates: its Role in Sustainable Agriculture.* CAB International, Wallingford, UK.

Guild, W.J. McL. 1952. Variation in earthworm numbers within field populations. *J. Anim. Ecol.* 21:169-181.

Guild, W.J. McL. 1955. Earthworms and soil structure. p 83-98. In: D.K. McE. Kevan, (ed.). *Soil Zoology*. Butterworth, London.

Hamblin, A.P. 1985. The influence of soil structure on water movement, crop root growth and water uptake. *Adv. Agronomy* 38:95-158.

Hamblin, A.P. 1991. Sustainable agricultural systems: what are the appropriate measures for soil structure? *Aust. J. Soil Res*. 29:709-715.

Hamblin, A.P. and D. Tenant. 1979. Interaction between soil type and tillage level in a dryland situation. *Aust. J. Soil Res.* 17:177-189.

Hole, F.D. 1981. Effects of animals on soil. *Geoderma* 25:75-112.

Hoogekamp, M.H., H. Rognar, and H.J.C. Eijsackers. 1983. Effects of earthworms on grassland and recently recliamed polder soils in the Netherlands. p85-105. In: J.E. Satchell (ed.), *Earthworm Ecology.* Chapman and Hall, London.

Hopp, H. and C.S. Slater. 1948. Influence of earthworms on soil productivity. *Soil Sci.* 66:421-428.

Jeanson, C. 1961. Sur une methode d'étude du comportement de la faune du sol et sa contribution a la pédogenèse. *C. R. Hebd. Seances Acad. Sci.* 253:2571-2573.

Jeanson, C. 1964. Micromorphology and experimental soil zoology; contribution to the study, by means of giant-sized thin sections, of earthworm-produced artificial soil structure. p.47-55. In: A. Jongerius (ed.) *Soil Micromorphology. Proc. 2nd Int. Wk Mtg Soil Micromorph.*, Arnhem.

Jeanson, C. 1971. Structure d'une galerie de lombric a la microsonde electronique.p.513-525. In: J. Jongerius (ed.) *IV Colloquium Pedobiologiae* I.N.R.A. Publ.71-7, Paris.

Kladivko, E.J. and H.J. Timmenga. 1990. Earthworms and agricultural management. p. 192-196. In: J.E. Box and L.C. Hammond (eds.) *Rhizosphere Dynamics*. Westview Press, Inc, Boulder, CO, USA.

Kobel-Lamparski, A. and F. Lamparski 1987. Burrow constructions during the development of *Lumbricus badensis* individuals. *Biol. Fertil. Soils* 3:125-129.

Kretzschmar, A. 1978. Quantification écologique des galeries de lombriciens. Techniques at premieres estimations. *Pedobiologia* 18:31-38.

Kretzschmar, A. 1982. Description des galeries de vers de terre et variation saisonnière des reseaux (observations en conditions naturelles). *Rev. Ecol. Biol. Sol* 19:579-591.

Kretzschmar, A. 1987a. Soil partitioning effect of an earthworm burrow system. *Biol. Fertil. Soils* 3:121-124.

Kretzschmar, A. 1987b. Caracterisation microscopique de 1 activite des lombriciens endogés. *Proc. 7th Int. Wkshp Soil Micromorph*, Paris p. 325-330.

Kretzschmar, A. and P. Monestiez. 1987. Topology, simulation and images of biological pores related to diffusion in soils. *Acta Stereol.* 6:355-360.

Krishnamoorthy, R.V. 1985. Nitrogen contribution by earthworm populations from grassland and woodland sites near Bangalore, India. *Rev. Ecol. Biol. Sol* 22:463-472.

Lavelle, P. 1983. The soil fauna of tropical savannas. II. The Earthworms. p.485-504.In: F. Bourlière (ed.) *Ecosystems of the World. 13. Tropical Savannas.* Elsevier, Amsterdam.

Lavelle, P., I. Barois, A. Martin, Z. Zaidi and R. Schaefer. 1989. Management of earthworm populations in agro-ecosystems: a possible way to maintain soil quality? p.109-22. In: M.Clarholm and L. Bergstrom (eds.) *Ecology of Arable Land.* Kluwer Academic Publishers, Stockholm.

Lee, K.E. 1959. The earthworm fauna of New Zealand. *N.Z. DSIR Bull.* 130:1-486.

Lee, K.E. 1983. The influence of earthworms and termites on soil nitrogen cycling p.35-48. In: Ph. Lebrun, H.M. André, A. de Medts, C. Gregroire-Wibo and G. Wauthy(eds.) *New Trends in Soil Biology.* Proc. 8th Int. Colloq. Soil Zoology,Louvain-la-Neuve 1982, Dieu-Brichart, Ottignies-Louvain-la-Neuve.

Lee, K.E. 1985. *Earthworms: their Ecology and Relationships with Soils and Land Use.* Academic Press, Sydney.

Lee, K.E. 1987. Peregrine species of earthworms. p. 315-27. In: A.M. Bonvicini and P.Omodeo (eds.) *On Earthworms*. UZI, Selected Symposia and Monographs No.2. Mucchi, Modena.

Lee, K.E. 1991. The diversity of soil organisms. p. 73-87. In: D.L. Hawksworth (ed.)*The Biodiversity of Microorganisms and Invertebrates: Its Role in Sustainable Agriculture*. CASAFA Report Series. CAB International, Wallingford, UK.

Lee, K.E. and R.C. Foster. 1991. Soil fauna and soil structure. *Aust. J. Soil Res.* 29:745-775.

Lee, K.E. and T.G. Wood. 1971a. Physical and chemical effects on soils of some Australian termites and their pedological significance. *Pedobiologia* 11:376-409.

Lee, K.E. and T.G. Wood. 1971b. *Termites and Soils*. Academic Press, London and NY.

Lobry de Bruyn, L.A. and A.J. Conacher. 1990. The role of termites and ants in soil modification. *Aust. J. Soil Res.* 28:55-93.

Logsdon, S.D. and R.R. Allmaras. 1991. Maize and soybean root clustering as indicated by root mapping. *Plant Soil*. 131:169-176.

Logsdon, S.D., R.R. Allmaras, W.W. Nelson and W.B. Voorhees. 1992. Persistence of subsoil compaction produced by heavy axle loads. *Soil Tillage Res.* 23:95-110.

Loquet, M. 1978. The study of respiratory and enzymatic activities of earthworms-made pedological structures in a grassland soil at Citeaux, France. *Scientif. Proc. R. Dublin Soc.* Ser. A6,207-214.

Moran, C.J., A.B. McBratney and A.J. Koppi. 1989. A rapid analysis method for soil macropore structure. I. Specimen preparation and digital binary image production. *Soil Sci. Soc. Am. J.* 53:921-928.

Moran, C.J. and A.B. McBratney. 1992. Acquisition and analysis of three-component digital images of soil pore structure. I. Method. *J. Soil Sci.* 43:541-549.

McKenzie, B.M. and A.R. Dexter. 1988a. Axial pressures generated by the earthworm *Aporrectodea rosea. Biol. Fertil. Soils* 5:323-327.

McKenzie, B.M. and A.R. Dexter. 1988b. Radial pressures generated by the earthworm *Aporrectodea rosea. Biol. Fertil. Soils* 5:328-332.

Nazaroff, P.S. 1931. Note on the spongy ironstone of Angola. *Geol. Mag.* 68:443-446.

Needham, A.E. 1957. Components of nitrogenous excreta in the earthworms *Lumbricus terrestris* L. and *Eisenia foetida* (Savigny). *J. Exp. Biol.* 34:425-446.

Nye, P.H. 1955. Some soil-forming processes in the humid tropics. IV. The action of the soil fauna. *J. Soil Sci.* 6:73-83.

Omoti, U. and A. Wild. 1979. Use of fluorescent dyes to mark the pathways of solute movement through soils under leaching conditions. 2. Field experiments. *Soil Sci.* 128:98-104.

Passioura, J.B. 1991. Soil structure and plant growth. *Aust. J. Soil Res.* 29:717-728.

Philip, J.R. 1985. Approximate analysis of the borehole permeameter in unsaturated soil. *Water Resour. Res.* 21:1025-1033.

Philip, J.R. 1987. An analog for infiltration and unsaturated seepage. *Eos.* 68:153-154.

Phillipson, J. and R.J. Bolton. 1977. Growth and cocoon production by *Allolobophora rosea* (Oligochaeta: Lumbricidae). *Pedobiologia* 17: 70-82.

Ratcliffe F.N. and T. Greaves. 1940. The subterranean foraging galleries of *Coptotermes lacteus* (Frogg.) *J. Coun. Sci. Ind. Res. Australia* 13:150-161.

Reddell, P. and A.V. Spain. 1991a. Earthworms as vectors of viable propagules of mycorrhizal fungi. *Soil Biol. Biochem.* 23:767-774.

Reddell, P. and A.V. Spain. 1991b. Transmission of infective *Frankia* (Actinomycetales) propagules in casts of the endogeic earthworm *Pontoscolex correthrurus*. *Zool. Anz.* 151:104-106.

Ringrose-Voase, A.J. 1991. Micromorphology of soil structure: description, quantification, application. *Aust. J. Soil Res.* 29:777-813.

Ross, D.J. and A. Cairns. 1982. Effects of earthworms and ryegrass on respiratory and enzyme activities of soil. *Soil Biol. Biochem.* 14:583-587.

Roth, C.H. and M. Joschko. 1991. A note on the reduction of runoff from crusted soils by earthworm burrows and artificial channels. *Z. Pflanzenernahr. Bodenk.* 154:101-105.

Rovira, A.D., K.R.J. Smettem and K.E. Lee. 1987. Effect of rotation and conservation tillage on earthworms in a red-brown earth under wheat. *Aust. J. Agric. Res.* 38:829-834.

Scharpenseel, H.W. and H. Gewehr. 1960. Studien zur wasserbewegung im beden mit tritium-wasser. *Z. Pfl. Ernahr. Dung.* 88:35-49.

Schnurer, J., M. Clarholm and T. Rosswall 1986. Fungi, bacteria and protozoa in soil from four arable cropping systems. *Biol. Fertil. Soils* 2:119-126.

Seymour, M. 1978. The infinite variety of worms. *New Scientist.* 77:650-652.

Shipitalo, M.J. and R. Protz. 1987. Comparison of morphology and porosity of a soil under conventional and zero tillage. *Can. J. Soil Sci.* 67:445-456.

Slater, C.S. and H. Hopp. 1947. Relation of fall protection to earthworm populations and soil physical conditions. *Soil Sci. Soc. Amer. Proc.* 12:508-511.

Sleeman, J.R. and R. Brewer. 1972. Micro-structures of some Australian termites. *Pedobiologia* 12:347-373.

Smettem, K.R.J. 1986. Analysis of water flow from cylindrical macropores. *Soil Sci. Soc. Am. J.* 50:1139-1142.

Smettem, K.R.J. 1987. Characterization of water entry into a soil with a contrasting textural class: spatial variability of infiltration parameters and influence of macroporosity. *Soil Sci.* 144:167-174.

Smettem, K.R.J. 1992. The relation of earthworms to soil hydraulic properties. *Soil Biol. Biochem.* 24:1539-1544.

Smettem, K.R.J. and N. Collis-George. 1985a. Statistical characterization of soil biopores using a soil peel method. *Geoderma.* 36:27-36.

Smettem, K.R.J. and N. Collis-George. 1985b. The influence of cylindrical macropores on steady-state infiltration in a soil under pasture. *J. Hydrol.* 79:107-114.

Smettem, K.R.J. and N. Collis-George. 1985c. Prediction of steady-state ponded infiltration distributions in a soil with vertical macropores. *J. Hydrol.* 70:115-122.

Smettem, K.R.J. and S.T. Trudgill. 1983. An evaluation of some fluorescent and non-fluorescent dyes in the identification of water transmission routes in soils. *J. Soil Sci.* 34:45-46.

Smettem, K.R.J. and P.J. Ross. 1992. Measurement and prediction of water movement in a field soil: the matrix-macropore dichotomy. *Hydrol. Processes.* 6:1-10.

Springett, J.A. 1983. Effect of five species of earthworm on some soil properties. *J. Appl. Ecol.* 20:865-872.

Springett, J.A. 1985. Effect of introducing *Allolobophora longa* Ude on root distribution and some soil properties in New Zealand pastures. p. 399-405. In: A.H. Fitter, D. Atkinson, D.J. Read and M.B. Usher (eds.) *Ecological Interactions in Soil.* Blackwell, Oxford.

Stockdill, S.M.J. 1966. The effect of earthworms on pastures. *Proc. N.Z. Ecol. Soc.* 13:68-75.

Stockdill, S.M.J. 1982. Effects of introduced earthworms on the productivity of New Zealand pastures. *Pedobiologia.* 24:29-35.

Syers, J.K. and J.A. Springett. 1984. Earthworms and soil fertility. *Plant Soil.* 76:93-104.

Teotia, S.P., F.P. Duley and T.M. McCalla. 1950. Effect of stubble mulching on number and activity of earthworms. *Nebraska Agric. Exp. Stn. Bull.* 165.

Tisdall, J.M. 1978. Ecology of earthworms in irrigated orchards. p 297-303. In: W.W. Emerson, R.D. Bond and A.R. Dexter (eds). *Modification of Soil Structure.* Wiley, Chichester, U.K.

Trojan, M.D. and D.R. Linden. 1992. Microrelief and rainfall effects on water and solute movement in earthworm burrows. *Soil Sci. Soc. Am. J.* 56:727-733.

Trueman, E.R. 1978. Locomotion. p 243-269. In: P.J. Mill (ed.). *Physiology of Annelids.* Academic Press, London.

Urbanek, J. and F. Dolezal. 1992. Review of some case studies on the abundance and on the hydraulic efficiency of earthworm channels in Czechoslovak soils with reference to the subsurface pipe drainage. *Soil Biol Biochem.* 24:1563-1571.

Van Stiphout, T.P.J., H.A.J. Van Lanen, O.H. Boersma, and J. Bouma. 1987. The effect of bypass flow and internal catchment of rain on the water regime in a clay loam grassland. *J. Hydrol.* 95:1-11.

Virginia, R.A., M.B. Jenkins and W.M. Jarrell. 1986. Depth of root symbiont occurrence in soil. *Biol. Fertil. Soils* 2:127-130.

White, I. and M.J. Sully. 1987. Macroscopic and microscopic capillary length and time scales from field infiltration. *Water Resour. Res.* 24:155-162.

White, I., M.J. Sully and K.M. Perroux. 1992. Measurement of surface-soil hydraulic properties: disk permeameters, tension infiltrometers and other techniques. In: Soil Sci. Soc. of Amer. (eds) *Advances in measurement of soil physical properties: bringing theory into practice.* SSSA Special Publication no 30., Madison, USA.

White, R.E. 1985. The influence of macropores on the transport of dissolved and suspended matter through soils. *Adv. Soil Sci.* 3:95-120.

Wiececk, C.S. and A.S. Messenger. 1972. Calcite contributions by earthworms to forest soils in northern Illinois. *Soil Sci. Soc. Am. Proc.* 36:478-480.

Wilke, D.E. von. 1953. Zur Kenntis der Lumbricidenfauna Deutschlands. *Zool. Anz.* 151:104-106.

Zicsi, A. 1983. Earthworm ecology in deciduous forests in central and southeast Europe. p.171-177. In: J.E. Satchell (ed.) *Earthworm Ecology.* Chapman and Hall, London.

Index